To my sons,
Joeby, Colin, and Kirk

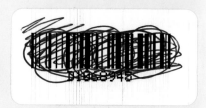

APPLICATIONS OF DISCRETE AND CONTINUOUS FOURIER ANALYSIS

APPLICATIONS OF DISCRETE AND CONTINUOUS FOURIER ANALYSIS

H. JOSEPH WEAVER

A Wiley-Interscience Publication

JOHN WILEY & SONS
New York · Chichester · Brisbane · Toronto · Singapore

Library of Congress Cataloging in Publication Data:
Weaver, H. Joseph.
    Applications of discrete and continuous Fourier analysis.

    "A Wiley-Interscience publication."
    Includes bibliographies and index.
    1. Fourier analysis.   I. Title.
QA403.5.W43   1983   515'.2433   83-3651
ISBN 0-471-87115-X

Printed in the United States of America

10 9 8 7 6 5 4 3

PREFACE

Fourier analysis is used to describe the behavior of a remarkably wide range of scientific and engineering phenomena. Since its introduction by Daniel Bernoulli over two centuries ago, it has been studied and developed to the point where it is perhaps the most useful form of mathematical analysis available today.

In this text, I have divided Fourier analysis into three analogous topics. First is the Fourier series—a mapping that converts a (periodic) function to a sequence of Fourier series coefficients. Next is the Fourier transform which maps a function to another function. Finally, we have the discrete Fourier transform which maps one sequence to another. Both the Fourier series and Fourier transform mappings have wide application in the study of physical phenomena. On the other hand, the utility of the discrete Fourier transform lies in the fact that it can be used to digitally calculate the other two mappings.

My intention, when designing this text, was to produce a work that would enable the reader to understand the concept and properties of the Fourier mappings and then apply this knowledge to the analysis of real scientific and engineering problems. To accomplish this goal, I have divided the text into two portions. The first, consisting of Chapters 1–5, presents the definition and properties of the Fourier series, Fourier transform, and discrete Fourier transform. Since this is a text on applications, the presentation in this first section does not dwell upon the questions of existence and reciprocity of the mappings but rather the emphasis is placed on how their properties can be used to simplify their calculation and manipulation. Chapters 6–11 constitute the second portion which builds upon the material presented in the first portion and demonstrates various physical applications of Fourier analysis.

In very simple terms, Fourier analysis is the study of how general functions can be decomposed into linear combinations of the trigonometric sine and cosine functions. Chapter 1 discusses this concept of frequency content of a function from both heuristic and mathematical points of view.

Chapters 2, 3, and 4 discuss the Fourier series, Fourier transform, and discrete Fourier transform, respectively. In each of these chapters the basic definition is presented and then the various properties are discussed. The presentations are presented in an analogous fashion so that the similarities of the mappings are highlighted.

Chapter 5 presents a discussion of the digital calculation of both the Fourier series and Fourier transform by using the discrete Fourier transform. In this chapter, sampling theory is discussed from a practical point of view.

Chapter 6 discusses the concepts of the impulse response and transfer function of a system. These concepts from systems theory are used throughout the remaining applications chapters.

The remaining five chapters deal directly with applications to specific fields. Specifically, Chapter 7 discusses application of Fourier analysis to both mechanical and electrical systems as well as the one-dimensional wave equation. Chapter 8 deals with the physics of optical wave propagation and optical systems engineering. Chapter 9 deals with the accuracy of numerical analysis algorithms from a frequency domain point of view. Chapter 10 discusses applications of Fourier analysis to the solution of the heat, or diffusion, equation. Chapter 11 discusses basic applications of Fourier analysis to statistics and probability theory as well as a brief presentation of stochastic systems analysis.

This text is based upon a portion of the material that was used as course notes at the University of California's Lawrence Livermore National Laboratory. This course was offered as part of their continuing education program. I am very grateful to several of those students whose comments and suggestions were very helpful in developing the new manuscript. In particular, my thanks to B. J. McKinley, Karena McKinley, Henry Chau, and Dr. Jeff Richardson for their fine efforts. Very special thanks must go to Dr. Gary Sommargren, a good friend and colleague, for his expert advice, criticism, and encouragement throughout the development of this text. Finally, deepest gratitude to my wife Sue who gave up many evenings to help type and proof this manuscript.

H. JOSEPH WEAVER

Livermore, California
May 1983

CONTENTS

APPLICATIONS OF DISCRETE AND CONTINUOUS FOURIER ANALYSIS

CHAPTER

1

THE CONCEPT OF FREQUENCY CONTENT

Fourier analysis, or frequency analysis, in the simplest sense is the study of the effects of adding together sine and cosine functions. This type of analysis has become an essential tool in the study of a remarkably large number of engineering and scientific problems. Daniel Bernoulli, while studying vibrations of a string in the 1750s, first suggested that a continuous function over the interval $(0, \pi)$ could be represented by an infinite series consisting only of sine functions. This suggestion was based on his physical intuition and was severely attacked by mathematicians of the day. Roughly 70 years later J. B. Fourier reopened the controversy while studying heat transfer. He argued, more formally, that a function continuous on an interval $(-\pi, \pi)$ could be represented as a linear combination of both sine and cosine functions. Still his conjecture was not readily accepted, and the question went unresolved for many years.

The purpose of this chapter is to present the concept of frequency content, or to say it another way, to study the effect of a linear combination of sine and cosine functions. To do this, we present both the definition and functional behavior of these basic building blocks (sine and cosine functions). At first, our approach is rather heuristic and the emphasis placed on the trigonometric derivation of these functions. From this derivation their functional behavior is demonstrated. Simple examples are used to illustrate how a combination of these functions is dependent upon both the amplitude and frequency of the

1

individual trigonometric function components. Once the "physical" concepts have been presented, we present a concise mathematical definition of frequency content. Finally, we show how complex variable theory can be used to enhance our understanding and appreciation of the concept of frequency content.

TRIGONOMETRIC SINE AND COSINE FUNCTIONS

The fundamental building blocks of Fourier analysis are the sine and cosine trigonometric functions. Trigonometry literally means "the measurement of three angles" and was first used to study the relationship between the sides and angles of a triangle. Today, however, the trigonometric functions themselves have become central objects of study in the modern mathematical field called analysis. The mathematician uses sines and cosines to study other functions, whereas the engineer and scientist use them to study certain periodic phenomenon. Although the sine and cosine functions can be introduced from either point of view, we approach the subject from the classical or "triangle solving" method. This serves to present the more physical side of these functions from which their periodic nature and analytical properties can be demonstrated. The right triangle shown in Figure 1.1 will be our starting point. The relations sine,

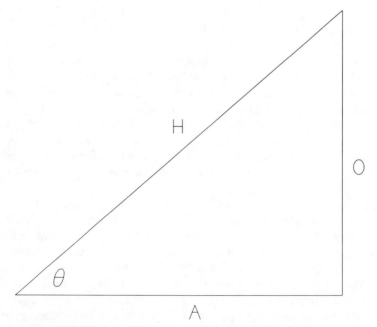

FIGURE 1.1. Trigonometric relations on a right triangle.

cosine, and tangent are defined on this triangle as follows:

$$(1.1a) \qquad \sin\theta = \frac{O}{H},$$

$$(1.1b) \qquad \cos\theta = \frac{A}{H},$$

$$(1.1c) \qquad \tan\theta = \frac{O}{A}.$$

Equations (1.1a)–(1.1c) are our basic definitions of the sine and cosine relations. These relations can be considered to give the measure of an angle in terms of the sides of a right triangle. The sine and cosine functions are dimensionless. The dimensions of the angle θ (which is discussed shortly) may be radians, gradians, degrees, or any other convenient measure one cares to define. Usually associated with trigonometry and the study of triangles are the units of degrees. However, we prefer radians. Let us now add to our repertoire the result of the well-known Pythagorean theorem which states

$$(1.2) \qquad H^2 = A^2 + O^2.$$

This equation can be used in conjunction with Equations (1.1) to derive many familiar results. For example, if we first solve Equations (1.1) for O and A in terms of sine and cosine and then square the resulting equations we obtain

$$(1.3) \qquad \begin{aligned} O^2 &= H^2\sin^2\theta, \\ A^2 &= H^2\cos^2\theta. \end{aligned}$$

Substitution of these equations into Equation (1.2) and dividing by H^2 yields

$$(1.4) \qquad \sin^2\theta + \cos^2\theta = 1.$$

As a second example, let us divide Equation (1.1a) by (1.1b) to arrive at another basic equation:

$$(1.5) \qquad \tan\theta = \frac{\sin\theta}{\cos\theta}.$$

Many other well-known results such as the law of sines and the law of cosines, as well as the sine and cosine of sums of angles can also be derived by similar trivial moves.

GENERALIZED SINE AND COSINE FUNCTIONS

When "triangle solving," it is really never necessary to consider angles greater than 90°. However, it is useful to generalize the sine and cosine functions to accommodate angles greater than this. The construction shown in Figure 1.2*a* helps to illustrate this generalization. Shown in the figure is a circle (of radius *R*) divided into four sections or quadrants. These quadrants are labeled from I to IV in the counterclockwise direction. The line drawn, at an angle θ, from the origin (center of circle) to the circumference is known as the radius vector. The angle θ is measured from the horizontal axis to the radius vector and is considered positive when swept in a counterclockwise direction.

We now construct the right triangle by dropping a line from the tip of the radius vector to the horizontal axis to form the opposite side (*O*) of the triangle. The value or "length" of this side is measured from the horizontal axis to the tip of the radius vector. A line drawn from the origin to the point of intersection of the *O* side and the horizontal axis gives the adjacent (*A*) side and completes the construction of the triangle. Again, the value or "length" of this side is measured from the origin and depending upon the angle θ will be positive or negative. The generalized sine and cosine functions of the angle θ are defined exactly as in Equations (1.1) only now we permit *O* and *A* to take on negative values. Note that the length of the radius vector is always considered positive. Depending upon the angle θ, we have four possible locations for the right triangle (in quadrants I through IV). For example, shown in Figure 1.2*b* is a value of θ that places the triangle in the second

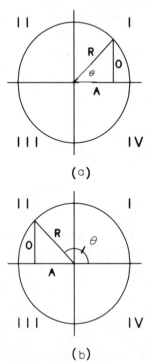

(a)

(b)　　**FIGURE 1.2.** Generalized trigonometric relations.

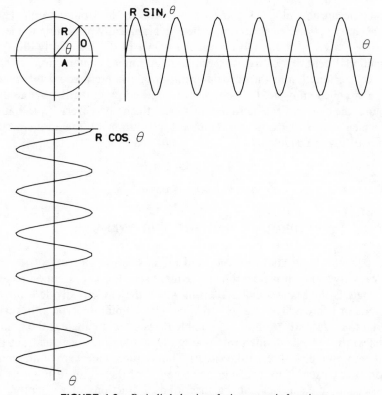

FIGURE 1.3. Periodic behavior of trigonometric functions.

quadrant. It is important to note that for these generalized sine and cosine functions the angle θ does not necessarily lie between the hypotenuse and the adjacent side of the triangle but is always measured from the positive horizontal axis to the radius vector in the counterclockwise direction.

The "triangle solving" equations still apply to the triangles regardless of the quadrant in which they are located. For example, in Figure 1.2b the angle between the hypotenuse and adjacent side is $180° - \theta$. Solving this triangle and comparing the results to the generalized trigonometric functions we obtain

$$\sin(180° - \theta) = \sin\theta,$$

(1.6)

$$\cos(180° - \theta) = -\cos\theta.$$

Let us now carry this construction one step farther to illustrate the functional behavior of the sine and cosine.[†] In Figure 1.3 we again use the circle

---

[†] From here on we drop the term generalized sine and cosine functions and simply refer to them as the sine and cosine functions.

construction but now plot $O\,(R\sin\theta)$ versus θ on the horizontal graph. This is a simple construction and is accomplished by graphically projecting the length of the side O over to the graph for various values of θ. This clearly illustrates the functional behavior of the sine function. A similar construction is shown for the cosine function in the vertical graph. Motion described by this type of function is called sinusoidal or simple harmonic motion. As is obvious from this figure, the sine and cosine functions repeat themselves every 2π radians or $360°$, which corresponds to one complete revolution of the radius vector. Mathematically, we have:

$$\sin(\theta + 2\pi) = \sin\theta,$$

(1.7)

$$\cos(\theta + 2\pi) = \cos\theta.$$

DEGREES, RADIANS, AND GRADS

We have been talking about degrees and radians as the measure of an angle rather casually so far. It is now time to be more specific. The common measure of the angle is the sexagesimal system in which the circle is divided into 360 basic units or degrees. The degree is divided into 60 minutes which, in turn, is divided into 60 seconds. A more reasonable measure of an angle is the grad. In this system the circle is divided into 400 grads or 100 grads per quadrant (this is an attempt at metrication of the circle). The rotational measure of an angle is also commonly used. If we consider one rotation of the vector in Figure 1.3 as our basic unit, then angles can be measured as fractions of a rotation. For example, $\frac{1}{5}$ rotation is equal to $72°$.

Although the degree, grad, and rotation systems are sufficient to measure an angle for triangle solving, the radial measure is much more useful when discussing analytical properties of the trigonometric functions. As is well known, if we were to take the radius of a circle, bend it into an arc, and lay it along the circumference, it would take $6.28319\ldots$, or 2π, radial lengths to completely fit around the circle. Thus we can reasonably ask, "Suppose we take a fraction of the radius (bend it) and lay it on the circumference as shown in Figure 1.4; what angle would it span?"

We now have a measure of the angle θ in terms of the length of a fraction (possibly greater than one) of the radius $A'B'$. In fact, the angle θ is equal to the length of the arc AB for a unit circle. This is a most important concept and is very useful to us in the next section.

DERIVATIVES OF THE SINE AND COSINE FUNCTIONS

In this section we take a look at the derivatives of the sine and cosine functions. In line with our stated philosophy we use triangles or the classical

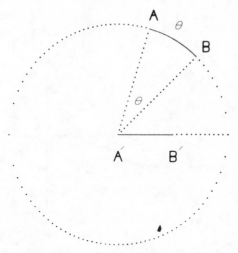

FIGURE 1.4.   Radial measure of an angle.

approach to derive some required preliminary results using the construction shown in Figure 1.5. (*Note*: The radius of the circle is one and thus we have a unit circle.) It is evident from the figure[†] that the length of the line PP' is less than the arc length PAP' which, in turn, is less than the length of the line RR'. Mathematically, we write

$$PP' \leqslant \text{arc } PAP' \leqslant RR'.$$

If we now divide each term of the previous inequality by two, and recall that the arclength of $PA = \theta$ when radial measure of an angle is used, we obtain

$$PC \leqslant \theta \leqslant RA.$$

Now, because the circle in the figure has radius 1, Equations (1.1) reduce to $PC = \sin \theta$ and $RA = \tan \theta$. Thus,

(1.8) $\sin \theta \leqslant \theta \leqslant \tan \theta.$

(It is important to note that this equation was derived using the radial measure of an angle and is not valid when other units such as degrees or grads are used.) Dividing relation (1.8) by $\sin \theta$ and then taking the reciprocal of each term we arrive at

(1.9) $\cos \theta \leqslant \dfrac{\sin \theta}{\theta} \leqslant 1.$

[†] The proof of this can be found in any elementary calculus text.

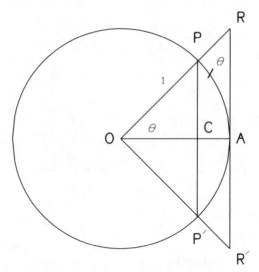

FIGURE 1.5. Similar triangles on the unit circle.

This relation helps us to establish two very important limits necessary to determine the derivatives of the sine and cosine functions. We note that $\cos \theta$ approaches 1 as θ goes to 0, and since $(\sin \theta)/\theta$ always lies between $\cos \theta$ and 1, it also must approach 1. That is,

$$(1.10) \qquad \lim_{\theta \to 0} \frac{\sin \theta}{\theta} = 1.$$

The expression $(\sin \theta)/\theta$ occurs so often that it is given its own name or symbol, $(\sin \theta)/\theta = \operatorname{sinc} \theta$. The second required limit is obtained as follows. From Equation (1.4) we have

$$\sin^2 \theta = 1 - \cos^2 \theta = (1 + \cos \theta)(1 - \cos \theta).$$

Now dividing both sides by $\theta(1 + \cos \theta)$ we obtain

$$\frac{\sin^2 \theta}{\theta(1 + \cos \theta)} = \frac{1 - \cos \theta}{\theta}.$$

Thus

$$\lim_{\theta \to 0} \frac{1 - \cos \theta}{\theta} = \lim_{\theta \to 0} \frac{\sin^2 \theta}{\theta(1 + \cos \theta)} = \left(\lim_{\theta \to 0} \frac{\sin \theta}{\theta} \right)\left(\lim_{\theta \to 0} \frac{\sin \theta}{1 + \cos \theta} \right) = (1)(0) = 0.$$

In summary,

$$(1.11) \qquad \lim_{\theta \to 0} \frac{1 - \cos \theta}{\theta} = 0.$$

We are now in a position to calculate the derivatives of the sine and cosine functions. Recall from elementary calculus the definition of the derivative of a function:

$$\frac{df(x)}{dx} = f'(x) = \lim_{\Delta x \to 0} \frac{f(x + \Delta x) - f(x)}{\Delta x}.$$

When applied to the sine function, the preceding equation becomes

$$\frac{d \sin \theta}{d\theta} = \lim_{\Delta \theta \to 0} \frac{\sin(\theta + \Delta \theta) - \sin \theta}{\Delta \theta}.$$

To determine this limit we first use the rule for the sine of the sum of two angles $[\sin(\alpha + \beta) = \sin \alpha \cos \beta + \sin \beta \cos \alpha]$.

$$\frac{d \sin \theta}{d\theta} = \lim_{\Delta \theta \to 0} \left(\frac{\sin \theta \cos \Delta \theta + \sin \Delta \theta \cos \theta - \sin \theta}{\Delta \theta} \right)$$

$$= \lim_{\Delta \theta \to 0} \sin \theta \left(\frac{\cos \Delta \theta - 1}{\Delta \theta} \right) + \lim_{\Delta \theta \to 0} \left(\frac{\sin \Delta \theta}{\Delta \theta} \right) \cos \theta.$$

Now using Equations (1.10) and (1.11) we have

$$(1.12) \qquad \frac{d \sin \theta}{d\theta} = \cos \theta.$$

Analogous reasoning yields the following:

$$(1.13) \qquad \frac{d \cos \theta}{d\theta} = - \sin \theta.$$

Inasmuch as the derivative of the sine is the cosine and the derivative of the cosine is the negative sine, we see that both the sine and cosine functions possess derivatives of all orders. In fact, the derivatives cycle every fourth time.

AMPLITUDE, FREQUENCY, AND PHASE

The functional behavior of $\sin x$ and $\cos x$ has been discussed and illustrated in Figure 1.3. These functions vary in a prescribed way between $+1$ and -1

and repeat themselves every 2π radians. Let us now consider $A \sin \mu x$ and $A \cos \mu x$ which are somewhat more general functions. The constant A is called the amplitude and μ is called the radial frequency. The amplitude is simply a constant that scales the "height" of the sine and cosine functions and causes them to now vary in the same prescribed way between $+A$ and $-A$. The radial frequency μ is a measure of how often the functions repeat themselves. Sin x repeats every 2π radians whereas $A \sin \mu x$ repeats every $2\pi/\mu$ radians. Since the argument of both the sine and cosine functions is in radians, the radial frequency μ has the dimensions of radians per x dimension. For example, if x represents time measured in seconds, then μ has the dimensions of radians per second. Radians per second is a difficult measure with which to physically identify. A more psychologically pleasing representation of frequency is the number of cycles or complete revolutions of the radius vector of Figure 1.3. When measured in cycles the frequency is called circular frequency and is denoted as ω. Obviously, $2\pi\omega = \mu$.

The period T of $A \sin 2\pi\omega x$, or $A \cos 2\pi\omega x$ is defined as the number of x units required to complete one cycle or 2π radians. It is given mathematically as $T = 1/\omega = 2\pi/\mu$. When the variable x represents a spatial measurement (such as the length of a vibrating string), a slightly different terminology is sometimes used. The period T is called the wavelength and is denoted by λ, whereas the radial frequency is called the wave number and is denoted by k.

Example

Consider the function $f(x) = 10 \sin 20x$, where x is distance measured in centimeters. Draw a graph of this function and determine the amplitude, period (wavelength), radial frequency (wave number), and circular frequency.

Solution. The graph of the function is shown in Figure 1.6. From the previous definitions we determine the amplitude to be 10, the period to be 0.3142 cm, the radial frequency to be 20 radians/cm, and the circular frequency to be 3.183 cycles/cm.

When $x = 0$ we have $\sin 2\pi\omega x = 0$ and $\cos 2\pi\omega x = 1$. It is often convenient to be able to shift the functions so that at $x = 0$ they may take on other values. This is accomplished by the addition of a "phase" term to the argument. For example, $\sin(2\pi\omega x - \varphi)$ is the same as $\sin 2\pi\omega x$ shifted to the "right" by an amount φ. This is shown in Figure 1.7.

CONCEPT OF FREQUENCY CONTENT

Fourier analysis, as demonstrated in the following chapters, is in effect the study of representing arbitrary functions by adding together various sine and

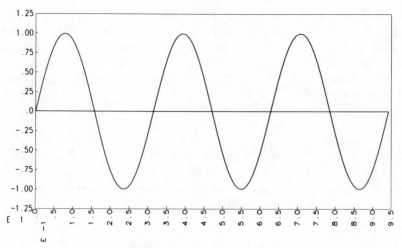

FIGURE 1.6.   Example problem.

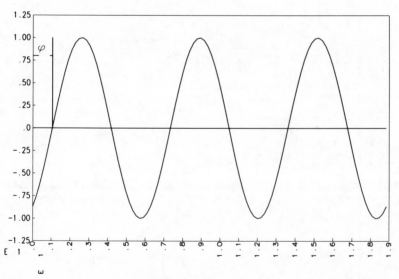

FIGURE 1.7.   Phase delay.

cosine functions. For a real appreciation of Fourier analysis it is necessary to understand the concept of frequency content of a function. The remainder of this chapter is devoted to presenting that concept.

In the previous section we saw that if we know both the amplitude and frequency of a sine or cosine function (without phase), then we know all about its functional behavior. That is, we can draw a graph of the function. This fact suggests that when several different sine and cosine terms are added together, the resulting function can be completely characterized by a knowledge of the amplitude and frequency of the sinusoidal terms involved. It is this concept that we explore, however, we first present a few simple illustrations of functions formed by adding sine terms together. Consider the function $f(t)$ made up of two sine terms of (circular) frequencies 10 and 15 cycles as per the following equation:

$$(1.14) \qquad f(t) = 3\sin 2\pi 10t + 2\sin 2\pi 15t$$

Shown in Figure 1.8a are the two individual sine terms and in Figure 1.8b, their sum $f(t)$. The important feature to note in this illustration is that $f(t)$ is not a trigonometric function but instead has its own unique behavior.

As a second illustration consider the function defined as

$$(1.15) \qquad g(t) = \sum_{k=1}^{5} B_k \sin 2\pi \omega_k t.$$

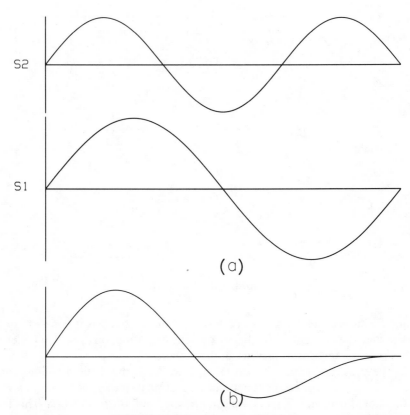

FIGURE 1.8.   Sum of two sine functions.

where $B_1 = 0.60,$ $\omega_1 = 1$
$B_2 = -0.20,$ $\omega_2 = 3$
$B_3 = 0.10,$ $\omega_3 = 5$
$B_4 = -0.09,$ $\omega_4 = 7$
$B_5 = 0.07,$ $\omega_5 = 9.$

The individual sine terms are shown in Figure 1.9a and their sum $g(t)$ in Figure 1.9b. Again, we note that $g(t)$ has its own unique shape and bears little resemblance to any of the sine functions in Figure 1.9a that constitute it. Yet it is important to observe that each of these sine functions contributes something to the shape of $g(t)$. This can be illustrated rather clearly by removing the second term in the summation, as shown in Figure 1.10, and comparing the results to that of Figure 1.9. From a physical or intuitive point of view, the higher (larger) frequency terms help to make up the finer details of the function. Whereas the lower (smaller) ones contribute more to the overall or basic shape of the function.

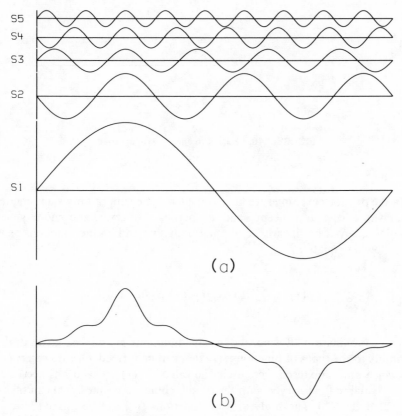

FIGURE 1.9. Sum of five sine terms.

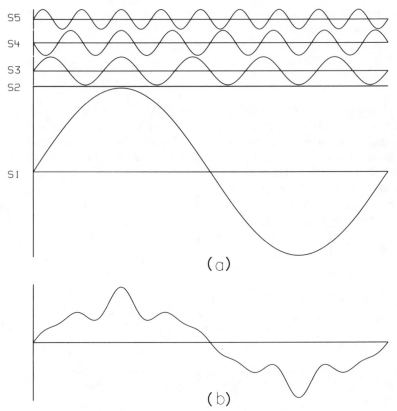

FIGURE 1.10.   Second frequency term removed.

In the previous examples we have considered the sum of only sine functions to illustrate certain concepts and properties. However, the more general situation is given by linearly combining both sine and cosine functions as per the following equation:

$$(1.16) \qquad f(t) = \sum_{k=1}^{n} A_k \cos 2\pi\omega_k t + B_k \sin 2\pi\omega_k t.$$

This combination of sine and cosine functions permits a great deal of flexibility in the types of functions $f(t)$ that can be formed. One obvious fact is that when only sine terms were used, the value of $f(t)$ at $t = 0$ was necessarily always equal to 0. However, with the proper choice of cosines, $f(t)$ can take on any value at $t = 0$. The study of what functions $f(t)$ can be formed by using Equation (1.16) is the cornerstone of Fourier series and is discussed in greater detail in Chapter 2. It is most important to note that $f(t)$ will have its own unique shape which, in general, will not resemble any of the trigonometric functions used to form it. However, each of the sine and cosine functions do contribute something to the overall shape of $f(t)$. As an example, consider

$$(1.17) \qquad f(t) = \sum_{k=1}^{3} A_k \cos 2\pi\omega_k t + B_k \sin 2\pi\omega_k t,$$

where $A_1 = 0$, $B_1 = 2$, $\omega_1 = 2$, $A_2 = -1$, $B_2 = 4$, $\omega_2 = 25$, $A_3 = 3$, $B_3 = 0$, and $\omega_3 = 40$. This is shown in Figure 1.11. The contribution of each sine and/or cosine term to the overall shape of $f(t)$ is obviously dependent on its amplitude. This can be rather dramatically illustrated by considering the function

$$f(t) = A_1 \cos 2\pi\omega_1 t + A_2 \cos 2\pi\omega_2 t + B_3 \sin 2\pi\omega_3 t,$$

where $\omega_1 = 1$
$\qquad \omega_2 = 4$
$\qquad \omega_3 = 10.$

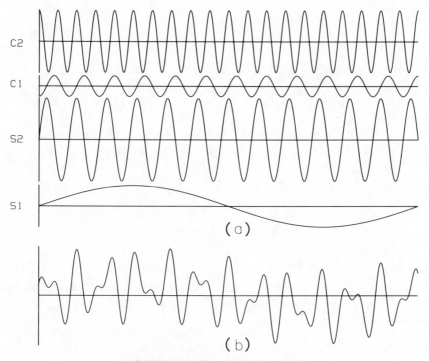

FIGURE 1.11. Equation (1.17) example.

The function $f(t)$ is shown in Figure 1.12 for $A_1 = 1$, $A_2 = 2$, and $B_3 = 3$. Figure 1.13 shows $f(t)$ for $A_1 = 2$, $A_2 = 3$, $B_3 = 1$, and Figure 1.14 shows $f(t)$ for $A_1 = 3$, $A_2 = 1$, $B_3 = 2$. The same three sine and cosine terms are present in each of the figures. However, their amplitudes have been interchanged. As can be seen, $f(t)$ has an entirely different shape in each of the figures. Thus we see the shape or functional behavior of $f(t)$ depends not only on which frequencies are present but also on "how much" of each frequency is present.

Up to this point we have been hinting at the meaning of frequency content of a function, but now the time has come to be more specific and formal. The frequency content of a function, described by Equation (1.16), is a measure of all the frequencies used in the summation (ω_k, $k = 1, n$), in which mode they are used (sine or cosine), and how much of each is used. Mathematically, we offer the following definition. Let f be a real function described by Equation (1.16). Then we say the *frequency content* of f is the set of 3-tuples (A_k, B_k, ω_k), $k = 1, n$. A_k is called the pure cosine content of frequency ω_k and, similarly, B_k is called the pure sine content of frequency ω_k.

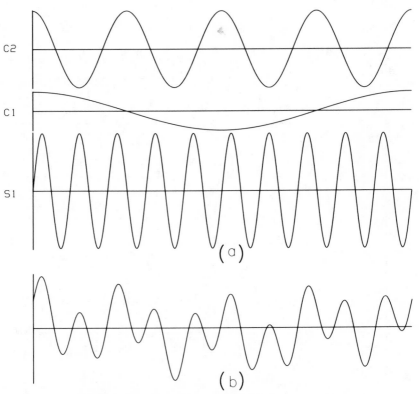

FIGURE 1.12.   Example problem ($A_1 = 1$, $A_2 = 2$, $B_3 = 3$).

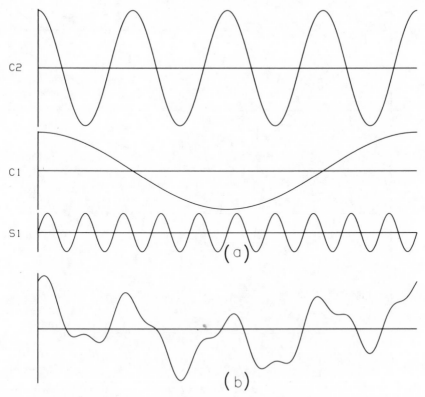

FIGURE 1.13.   Example problem ($A_1 = 2$, $A_2 = 3$, $B_3 = 1$).

Now let us step back and take this definition apart. Mathematically, a set is nothing more than a collection of objects. In our case these objects are 3-tuples. A 3-tuple is simply three real numbers arranged in a specific order. That is to say, it requires three real numbers to describe a 3-tuple and the order in which these numbers are specified is important [i.e., (A_k, B_k, ω_k) is not necessarily equal to (B_k, ω_k, A_k), etc.]. A common example of a 3-tuple is a three-dimensional vector.

A knowledge of the frequency content of a function is equivalent to a knowledge of the amplitude and frequency of all the terms in the summation of Equation (1.16). In reality, the frequency content of a function is a convenient or compact notational representation of equation (1.16). As an example, the set

$$\{(0, 2, 2), (-1, 4, 25), (3, 0, 40)\}$$

is the frequency content of the function given by Equation (1.17) (shown in Figure 1.11).

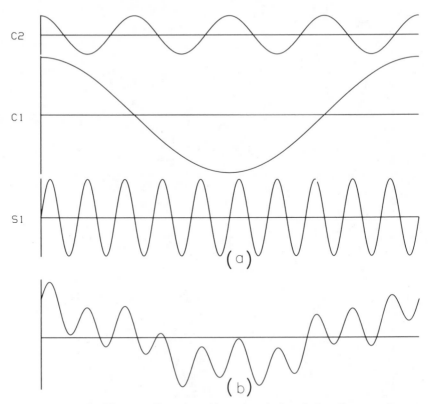

FIGURE 1.14.   Example problem ($A_1 = 3$, $A_2 = 1$, $B_3 = 2$).

The most comfortable way to "read" the frequency content of a function is to construct graphs of A_k and B_k versus ω_k. These graphs are called frequency domain plots. More specifically, the graph of A_k versus ω_k is called the pure cosine frequency plot and B_k versus ω_k is the pure sine frequency plot. For example, the frequency plots for the function given by Equation (1.17) are shown in Figure 1.15.

These frequency plots turn out to be extremely valuable tools in Fourier analysis. Again, we stress the fact that these graphs are merely a convenient way to represent the frequency content of a function which, in turn, is a compact notational representation of the summation of Equation (1.16).

It is to our advantage to discuss the techniques of reading and constructing these frequency plots. However, inasmuch as these plots usually appear in a somewhat more general or complex form, we must first briefly digress and touch base with some elementary complex variable theory.

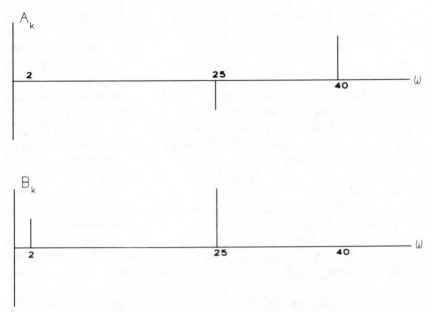

FIGURE 1.15. Frequency plots of Equation (1.17).

COMPLEX VARIABLE TUTORIAL

In this section we lightly go over some results from basic complex variable theory. These results are needed for much of the material covered in the remainder of this book. We use a relatively rigorous mathematical approach that permits us to present all the basic results in a very clean and compact form. This type of approach also gives us a solid foundation from which to study the more familiar rectangular and polar representation of a complex number. We begin with the following definition.

A COMPLEX NUMBER is an ordered pair of real numbers that we denote as (x, y). The first number x is called the real part of the complex number and the second number y is called the imaginary part. Two complex numbers $z_1 = (x_1, y_1)$ and $z_2 = (x_2, y_2)$ are called equal (and we write $z_1 = z_2$) if and only if $x_1 = x_2$ and $y_1 = y_2$. We define the sum $(z_1 + z_2)$ and the product $(z_1 z_2)$ by the equations

$$z_1 + z_2 = (x_1 + x_2, y_1 + y_2),$$
$$z_1 z_2 = (x_1 x_2 - y_1 y_2, x_1 y_2 + y_1 x_2).$$

This is an extremely powerful and general definition of a complex number. Before we use it to obtain useful results, let us look at it more closely. It first says a complex number is an ordered pair of real numbers. That is to say, two real numbers (x and y) are required to describe one complex number, and these real numbers are ordered [i.e., (x, y) is not necessarily equal to (y, x)]. Since these numbers are ordered it makes sense to give them names (real and imaginary parts). This definition also spells out exactly what is meant by equality, addition, and multiplication of two complex numbers. Two complex numbers are equal if and only if their real parts are equal and their imaginary parts are equal. For example, $z_1 = (1, 2)$ and $z_2 = (2, 1)$ are not equal. To add two complex numbers together, we add their real parts together and add their imaginary parts together. For example, if $z_1 = (1, 2)$ and $z_2 = (4, 7)$, then $z_1 + z_2 = (5, 9)$. Based on the definitions of equality and addition we might expect that the definition of multiplication would require that we multiply the real parts and the imaginary parts together. It turns out, however, that this is not a useful definition of multiplication, so instead we have the afore-mentioned algorithm which mixes the real and imaginary parts. For example, if $z_1 = (1, 2)$ and $z_2 = (2, 1)$, then $z_1 z_2 = (0, 5)$.

Using the preceding definition of a complex number and the properties of real numbers, we can show that complex numbers obey the commutative, distributive, and associative laws. That is for $z_1 = (x_1, y_1)$, $z_2 = (x_2, y_2)$, and $z_3 = (x_3, y_3)$ we have the following:

Commutative law

$$z_1 + z_2 = z_2 + z_1$$

$$z_1 z_2 = z_2 z_1.$$

Associative law

$$z_1 + (z_2 + z_3) = (z_1 + z_2) + z_3$$

$$z_1(z_2 z_3) = (z_1 z_2)z_3.$$

Distributive law

$$z_1(z_2 + z_3) = z_1 z_2 + z_1 z_3.$$

Although we don't wish to become entrenched with an excess of mathematical manipulations, just for the sake of analytical clarification, we demonstrate the proof of the distributive law.

$$z_1(z_2 + z_3) = (x_1, y_1)(x_2 + x_3, y_2 + y_3)$$

$$= [x_1(x_2 + x_3) - y_1(y_2 + y_3), x_1(y_2 + y_3) + y_1(x_2 + x_3)]$$

$$= (x_1x_2 - y_1y_2, x_1y_2 + y_1x_2) + (x_1x_3 - y_1y_3, x_1y_3 + y_1x_3)$$

$$= z_1z_2 + z_1z_3.$$

The proofs of the other two laws are accomplished in an analogous fashion. Let us now consider a few special complex numbers.

The zero element (denoted as $\mathbf{0}$)[†] is the complex number $(0,0)$. This element has special properties. First, the sum of any complex number z and the zero element is z; that is,

$$z + \mathbf{0} = (x, y) + (0,0) = (x + 0, y + 0) = (x, y) = z.$$

Second, the product of any complex number and the zero element is the zero element; that is,

$$z\mathbf{0} = (x, y)(0,0) = (x0 - y0, x0 + y0) = (0,0) = \mathbf{0}.$$

The unit element (denoted as $\mathbf{1}$) is the complex number $(1,0)$ and has the property that the product of any complex number z and the unit element is z; that is,

$$z\mathbf{1} = (x, y)(1,0) = (x1 - y0, 1y + 0x) = (x, y) = z.$$

The imaginary unit element (denoted as i) is defined as the complex number $(0, 1)$. This element provides us with a solution to the equation $z^2 = -1$; that is,

$$z^2 = (0, 1)(0, 1) = (00 - 11, 01 + 10) = (-1, 0) = -1.$$

Given any complex number $z = (x, y)$, the negative of z (denoted as $-z$) is defined as the complex number $(-x, -y)$. Obviously, we have the following property:

$$z + (-z) = (x, y) + (-x, -y) = (x - x, y - y) = (0,0) = \mathbf{0}.$$

[†] When initially defining the complex unit and zero elements we use boldface notation. However, from here on we drop this boldface unless its omission would lead to confusion in distinguishing between the complex numbers 0 and 1, and the real numbers 0 and 1.

Given any complex number $z = (x, y)$, not equal to $(0,0)$, the inverse of z (denoted as z^{-1}) is defined to be the complex number,

$$\cdot \left(\frac{x}{x^2 + y^2}, \frac{-y}{x^2 + y^2} \right).$$

This inverse element has the property

$$zz^{-1} = (x, y)\left(\frac{x}{x^2 + y^2}, \frac{-y}{x^2 + y^2} \right)$$

$$= \left(\frac{x^2}{x^2 + y^2} + \frac{y^2}{x^2 + y^2}, \frac{-xy}{x^2 + y^2} + \frac{xy}{x^2 + y^2} \right)$$

$$= (1, 0) = \mathbf{1}.$$

Using the negative and inverse elements we are able to define subtraction and division of complex numbers. Subtraction of two complex numbers z_1 and z_2 (denoted as $z_1 - z_2$) is defined by

$$z_1 - z_2 = z_1 + (-z_2).$$

Similarly, division of the complex number z_1 by z_2, not equal to $(0, 0)$ (denoted as z_1/z_2), is defined by

$$\frac{z_1}{z_2} = z_1 z_2^{-1}.$$

Let us now consider three additional results that are very useful to us later. A real constant a can be written as the complex number $(a, 0)$. Thus the multiplication of any complex number z by a real constant a results in

$$az = (a, 0)(x, y) = (ax - 0y, 0x + ay) = (ax, ay).$$

A term that frequently occurs in complex analysis is $1/i$. This can be rationalized as follows:

$$\frac{1}{i} = i^{-1} = \left(\frac{0}{0 + 1}, \frac{-1}{0 + 1} \right) = (0, -1) = -1(0, 1) = -i.$$

We define the *modulus*, or absolute value, of a complex number $z = (x, y)$ to be the non-negative real number given by

$$\|z\| = \left(x^2 + y^2 \right)^{1/2}.$$

The complex conjugate of a complex number $z = (x, y)$ is defined as $z^* = (x, -y)$.

We are now able to justify expressing a complex number in the popular rectangular form. We present this result in the form of a theorem.

Theorem 1.1 (Rectangular Form of Complex Number). Every complex number $z = (x, y)$ can be represented in the form $x + iy$.

Proof.

$$x = (x, 0),$$

$$iy = (0, 1)(y, 0) = (0y - 10, 00 + y1) = (0, y).$$

Thus

$$x + iy = (x, 0) + (0, y) = (x, y).$$

This rectangular form is a natural for a geometric, or vector, interpretation of a complex number. In Figure 1.16 we have drawn the complex number $z = x + iy$, where x is laid out along the horizontal or real axis and y along the vertical or imaginary axis.

If we denote the length, or magnitude, of the vector z as r and its angle as θ (measured counterclockwise from the real axis), then we can use Equations (1.1) and (1.2) to write

$$x = r \cos \theta,$$

$$y = r \sin \theta,$$

$$r^2 = x^2 + y^2,$$

$$\tan \theta = \frac{y}{x}.$$

Using the preceding equations, z can be written in polar form,

(1.18) $$z = x + iy = r(\cos \theta + i \sin \theta).$$

The complex conjugate z^* is graphically illustrated in Figure 1.17. As can be seen from this figure, if z forms any angle θ with the real axis, then z^* will form an angle $-\theta$ with the real axis. That is, mathematically,

$$z = x + iy = r(\cos \theta + i \sin \theta)$$

$$z^* = x - iy = r[\cos(-\theta) + i \sin(-\theta)].$$

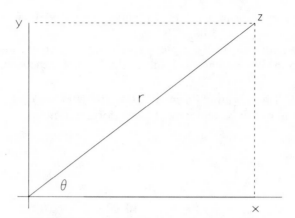

FIGURE 1.16. Rectangular form of complex number.

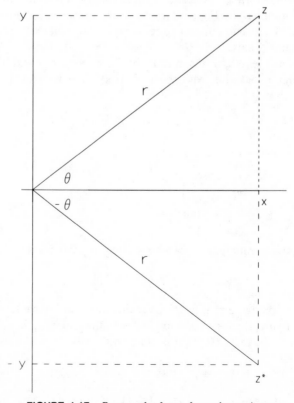

FIGURE 1.17. Rectangular form of complex conjugate.

Equation (1.18) can be placed in a more compact form known as the exponential form. To do this we must first derive Euler's equations. Using a Taylor series expansion we can write

$$(1.19) \qquad \sin x = x - \frac{x^3}{3!} + \frac{x^5}{5!} - \frac{x^7}{7!} + \cdots,$$

$$(1.20) \qquad \cos x = 1 - \frac{x^2}{2!} + \frac{x^4}{4!} - \frac{x^6}{6!} + \cdots,$$

and

$$(1.21) \qquad e^\theta = 1 + \theta + \frac{\theta^2}{2!} + \frac{\theta^3}{3!} + \frac{\theta^4}{4!} + \frac{\theta^5}{5!} + \cdots.$$

If we now substitute $i\theta$ for θ in Equation (1.21), we obtain

$$e^{i\theta} = 1 + i\theta - \frac{\theta^2}{2!} - \frac{i\theta^3}{3!} + \frac{\theta^4}{4!} + \frac{i\theta^5}{5!} + \cdots$$

$$= \left(1 - \frac{\theta^2}{2!} + \frac{\theta^4}{4!} + \cdots\right) + i\left(\theta - \frac{\theta^3}{3!} + \frac{\theta^5}{5!} + \cdots\right).$$

Now using Equations (1.19) and (1.20) we have

$$(1.22) \qquad e^{i\theta} = \cos\theta + i\sin\theta,$$

from which we also obtain

$$(1.23) \qquad e^{-i\theta} = \cos\theta - i\sin\theta.$$

Together Equations (1.22) and (1.23) are known as Euler's equations and can be used to rewrite equation (1.18) as

$$(1.24) \qquad z = re^{i\theta} = r\exp(i\theta),$$

and

$$(1.25) \qquad z^* = re^{-i\theta} = r\exp(-i\theta).$$

The term $\exp[i\theta]$ may be regarded as a unit vector rotated counterclockwise an angle θ from the real axis. Its real part is $\cos\theta$ and its imaginary part is $\sin\theta$. Using the definition of the modulus of a complex number we find that its magnitude is always unity:

$$\|e^{i\theta}\| = (\cos^2\theta + \sin^2\theta)^{1/2} = 1.$$

Euler's equations can be used to obtain a virtually unlimited number of useful results or identities. For example, addition and subtraction yield the following important equations:

$$(1.26) \qquad \cos\theta = \frac{e^{i\theta} + e^{-i\theta}}{2},$$

$$(1.27) \qquad \sin\theta = \frac{e^{i\theta} - e^{-i\theta}}{2i}.$$

Euler's equations afford a simple technique for multiplication and division of two complex numbers. That is, if $z_1 = r_1\exp(i\theta_1)$ and $z_2 = r_2\exp(i\theta_2)$ (r_2 not equal to 0), then

$$z_1 z_2 = r_1 r_2 e^{i(\theta_1 + \theta_2)},$$

and

$$\frac{z_1}{z_2} = \frac{r_1}{r_2} e^{i(\theta_1 - \theta_2)}.$$

Thus we see to multiply two complex numbers together in the exponential form we simply multiply their magnitudes and add their angles. Similarly, to divide we divide their magnitudes and subtract their angles.

COMPLEX REPRESENTATION OF FREQUENCY CONTENT

We are now in a position to discuss frequency content and frequency domain plots using complex variable theory. We stress the fact that this is simply a convenient mathematical tool that permits some additional insights into the problem. No new information is contained in this approach. We begin by using Equations (1.26) and (1.27) to rewrite Equation (1.16) in its complex form:

$$f(t) = \sum_{k=1}^{n} \left[\frac{A_k}{2} \left(e^{2\pi i \omega_k t} + e^{-2\pi i \omega_k t} \right) + \frac{B_k}{i2} \left(e^{2\pi i \omega_k t} - e^{-2\pi i \omega_k t} \right) \right]$$

$$(1.28)$$

$$f(t) = \sum_{k=1}^{n} \left(C_k e^{2\pi i \omega_k t} + C_{-k} e^{-2\pi i \omega_k t} \right)$$

where

$$C_k = \frac{A_k - iB_k}{2},$$

$$C_{-k} = \frac{A_k + iB_k}{2}.$$

Equation (1.28) is the complex representation of Equation (1.16). The term $\exp(2\pi i\omega_k t)$ is a rotating unit vector of circular frequency ω_k, whereas $\exp(-2\pi i\omega_k t)$ is a rotating unit vector of circular frequency $-\omega_k$. These can be thought of as rotating in opposite directions. The coefficients C_k and C_{-k} are complex numbers and, in fact, complex conjugates of each other. It is more convenient to talk about Equation (1.28) if we place it in a more compact notational form:

(1.29)
$$f(t) = \sum_{k=-n}^{n} C_k e^{2\pi i\omega_k t}$$

where

$$C_k = \frac{A_k + iB_k}{2}, \qquad k < 0,$$

$$C_0 = 0,$$

$$C_k = \frac{A_k - iB_k}{2}, \qquad k > 0,$$

$$\omega_k = -\omega_k, \qquad k < 0.$$

A plot of C_k versus ω_k ($k = -n, \ldots, -1, 0, 1, \ldots, n$) is called the complex frequency domain plot. Since each C_k is a complex number we require two graphs. The real part of the C_k terms contains only A_k and thus information concerning the pure cosine frequency content. Similarly, the imaginary part of the C_k terms contains information about the pure sine frequency content. For real functions there is no physical significance to the negative frequency terms. They are simply a result of a mathematical identity. For example, we could get negative frequencies by making use of the following properties of the trigonometric function:

(1.30)
$$A \cos 2\pi\omega t = A \cos(-2\pi\omega t),$$
$$B \sin 2\pi\omega t = -B \sin(-2\pi\omega t).$$

For illustrative purposes, the complex frequency domain plot for the function of Equation (1.17) is now presented. From previous work we know the frequency content of this function is given by the set

$$\{(0,2,2), (-1,4,25), (3,0,40)\}.$$

Using this set in conjunction with Equations (1.29) we have

k	C_k	ω_k
-3	$\frac{3}{2}$	-40
-2	$-\frac{1}{2} + 2i$	-25
-1	i	-2
1	$-i$	2
2	$-\frac{1}{2} - 2i$	25
3	$\frac{3}{2}$	40

This is shown in Figure 1.18 in rectangular form. The polar or exponential

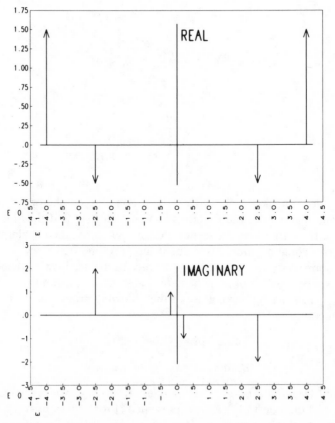

FIGURE 1.18. Frequency domain plot of Equation (1.23) (rectangular form).

form of the complex frequency domain plot is also commonly used. In this case we write C_k as

$$C_k = r_k e^{i\varphi_k}, \quad \text{for all } k,$$

$$r_k = \left(\frac{A_k^2 + B_k^2}{4} \right)^{1/2}$$

(1.31)

$$\tan \varphi_k = \frac{B_k}{A_k}, \qquad k < 0,$$

$$\tan \varphi_k = \frac{-B_k}{A_k}, \qquad k > 0.$$

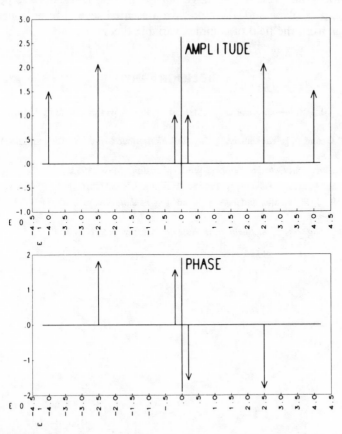

FIGURE 1.19. Frequency domain plot of Equation (1.23) (exponential form).

When placed in this form we have

$$C_k e^{2\pi i \omega_k t} = r_k e^{i\varphi_k} e^{2\pi i \omega_k t} = r_k e^{i(2\pi \omega_k t + \varphi_k)};$$

φ_k is thus referred to as the phase. Shown in Figure 1.19 is the exponential form of the frequency plot of the function of Equation (1.17).

SUMMARY

In this chapter we presented several fundamental mathematical concepts useful to us in our study of Fourier analysis. We first presented a basic definition and derived several analytical properties of the trigonometric sine and cosine functions. These functions are the basic building blocks of Fourier analysis. Next the concept of frequency content, or the linear combination of various sine and cosine functions, was examined and a reasonably concise mathematical definition was presented. A large portion of Fourier analysis is performed in the complex plane and, therefore, we presented several basic concepts and definitions from the field of complex variable theory.

BIBLIOGRAPHY

Apostal, T. M., *Mathematical Analysis*, 1st ed. Addison-Wesley, Reading, Mass., 1957, 4th printing.

Johnson, R. C. and F. L. Kiokemeister, *Calculus with Analytic Geometry*. Allyn and Bacon, Boston, 1960.

Kaplan, W., *Advanced Calculus*. Addison-Wesley, Reading, Mass., 1952.

Mostert, P. S., *Analytic Trigonometry*. Prentice-Hall, Englewood Cliffs, N.J., 1960.

Haaser, N. B., J. P. LaSalle, and J. A. Sullivan, *Intermediate Analysis*. Blaisdell Publishing, New York, 1964.

Raven, F. H., *Mathematics of Engineering Systems*. McGraw-Hill, New York, 1966.

CHAPTER

2

THE FOURIER SERIES

In Chapter 1 we discussed the concept of frequency content of a function or, in other words, the effect of adding together sine and cosine terms of various amplitudes and frequencies. The purpose of Chapter 1 was to introduce the reader to the idea of adding these trigonometric functions together. We had no specific formula for the A_k, B_k, and ω_k terms. In fact, they were usually chosen quite arbitrarily and then added together to illustrate some particular property or concept. In this chapter we are given some function $f(t)$ and required to find the A_k, B_k, and ω_k terms such that when they are added together they equal $f(t)$. In Chapter 1 we considered (but never required) the frequency content of a function to be a finite set. In this chapter we consider a very special type of frequency content, known as Fourier series frequency content which, in general, is an infinite set (a set with an infinite number of elements). Therefore, the problem in this chapter is: given a function $f(t)$, can we find real numbers A_k, B_k, and ω_k such that

(2.1) $$f(t) = \sum_{k}^{\infty} A_k \cos 2\pi\omega_k t + B_k \sin 2\pi\omega_k t.$$

or, in terms of the previous chapter, does $f(t)$ have frequency content? To solve this problem we must first find formulas for calculating A_k, B_k, and ω_k. Then we must show that when these terms are calculated according to the formulas and added together as in Equation (2.1), they will converge and more specifically converge to $f(t)$. In general, given any arbitrary function f, we can not guarantee that the series of Equation (2.1) will converge to f or that the A_k, B_k, and ω_k terms will even exist. In this chapter we cleverly avoid the problems of existence and convergence by proclaiming that *we only consider functions that possess a Fourier series* inasmuch as our purpose is to present the formulas for calculating the Fourier series frequency content of a function and to demonstrate how to manipulate these formulas. However, for the sake of completeness we now simply state the *Dirichlet conditions* which, if satisfied, guarantee that a function f will have Fourier series frequency content. The Dirichlet conditions are:

1. f is periodic with period T; that is, $f(t + T) = f(t)$.
2. f is bounded.
3. In any one period the function may have at most a finite number of discontinuities and a finite number of maxima and minima.

These are sufficient conditions but not necessary ones. That is to say, if these conditions are satisfied, then the function has Fourier series frequency content, but there are still other functions, that may not satisfy the conditions that also have Fourier series frequency content.

FOURIER SERIES FREQUENCY CONTENT

In the remainder of this chapter we assume that the Dirichlet conditions are satisfied for all functions considered. We now derive the formulas for A_k (pure cosine), B_k (pure sine), and ω_k (frequency) such that Equation (2.1) is valid. The frequency ω_k is given as

$$(2.2) \qquad \omega_k = \frac{k}{T}, \qquad k = 0, 1, 2, \ldots,$$

NOTE: We have changed our index k from $k = 1, 2, \ldots,$ to $k = 0, 1, \ldots.$ This permits us the use of an additional zero frequency or DC term.

The pure cosine frequency terms are given by

$$A_k = \frac{2}{T} \int_{-T/2}^{T/2} f(t) \cos \frac{2\pi kt}{T} \, dt, \qquad k = 1, 2, \ldots,$$

$$(2.3)$$

$$A_0 = \frac{1}{T} \int_{-T/2}^{T/2} f(t) \, dt.$$

The pure sine frequency content terms are given by

$$B_k = \frac{2}{T} \int_{-T/2}^{T/2} f(t) \sin \frac{2\pi kt}{T} dt, \qquad k = 1, 2, \ldots,$$

(2.4)

$$B_0 = 0.$$

When the constants A_k, B_k, and ω_k are given by Equations (2.2)–(2.4) we say the function f given by Equation (2.1) has Fourier series frequency content, or that Equation (2.1) is the Fourier series representation of f. This is a very special case of frequency content in which the frequencies are all multiples of a basic frequency $w_0 = 1/T$.

We now show how Equations (2.3) and (2.4) are obtained. To do this we must first present the following "orthogonality" relations:

(2.5) $$\int_{-T/2}^{T/2} \cos \frac{2\pi nt}{T} \cos \frac{2\pi mt}{T} dt = \begin{cases} 0, & n \neq m, \\ T/2, & n = m \neq 0, \end{cases}$$

(2.6) $$\int_{-T/2}^{T/2} \sin \frac{2\pi nt}{T} \sin \frac{2\pi mt}{T} dt = \begin{cases} 0, & n \neq m, \\ T/2, & n = m \neq 0, \end{cases}$$

(2.7) $$\int_{-T/2}^{T/2} \sin \frac{2\pi nt}{T} \cos \frac{2\pi mt}{T} dt = 0.$$

To obtain Equation (2.3) we first multiply both sides of Equation (2.1) by $\cos(2\pi mt/T)$ and then integrate the resulting expression from $-T/2$ to $T/2$:

$$\int_{-T/2}^{T/2} f(t) \cos \frac{2\pi mt}{T} dt = \int_{-T/2}^{T/2} \left(\sum_{n=1}^{\infty} A_n \cos \frac{2\pi nt}{T} \cos \frac{2\pi mt}{T} \right.$$

$$\left. + B_n \sin \frac{2\pi nt}{T} \cos \frac{2\pi mt}{T} \right) dt, \qquad m = 1, 2, \ldots.$$

Now using the orthogonality relations [Equations (2.5)–(2.7)] we see that for $m = n$ we have

$$\int_{-T/2}^{T/2} f(t) \cos \frac{2\pi mt}{T} dt = \frac{A_m T}{2} \quad \text{for } m = 1, 2, \ldots.$$

We have a special situation when $n = m = 0$. In this case,

$$\cos \frac{2\pi nt}{T} = \cos \frac{2\pi mt}{T} = 1,$$

$$\sin \frac{2\pi nt}{T} = 0,$$

and we obtain

$$\int_{-T/2}^{T/2} f(t)\, dt = \int_{-T/2}^{T/2} A_0\, dt = T A_0.$$

Equation (2.4) is obtained in a similar way; that is, multiply both sides of Equation (2.1) by $\sin(2\pi m t/T)$ and then integrate both sides of the resulting equation from $-T/2$ to $T/2$.

PERIODIC FUNCTIONS

The first Dirichlet condition requires that our function be periodic. However, there are many times when we are interested in the Fourier series of a function that is not periodic. In the real world we are usually interested in functions over a finite domain and do not care what the function does outside this domain. For example, we may want to determine the Fourier series of a function such as the one shown in Figure 2.1 over the interval $1 \leqslant t \leqslant 3$. Thus we are actually interested in the frequency content of the function shown in Figure 2.2. To solve this problem we simply extract that portion of f between 1 and 3 and generate a new, or *protracted*, periodic function from this extracted portion as illustrated in Figure 2.3. We then apply Equations (2.2)–(2.4) to this protracted function. There are problems that can arise when we extend a function in this way. The most obvious is the possibility of introducing discontinuities into the protracted function. For example, consider the function $f(t) = t^2$ over the closed interval [0, 2] as shown in Figure 2.4. This function is obviously continuous over this interval but its protracted function, shown in Figure 2.5, is discontinuous at $0, 2, 4, \ldots$. When extending a function, these discontinuities should be avoided if at all possible because they lead to Gibbs

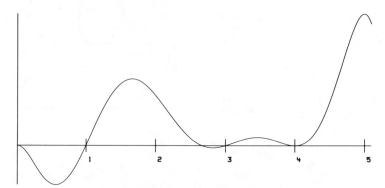

FIGURE 2.1.   Function for $t \geqslant 0$.

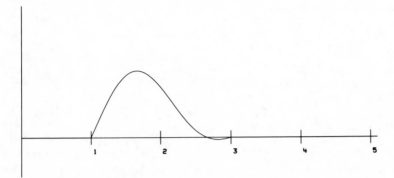

FIGURE 2.2. Function for $1 \leqslant t \leqslant 3$.

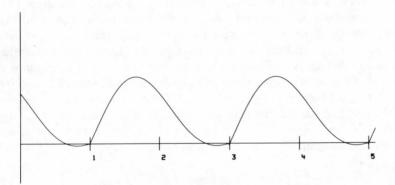

FIGURE 2.3. Protracted periodic function.

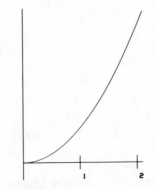

FIGURE 2.4. Function $f(t) = t^2$.

35

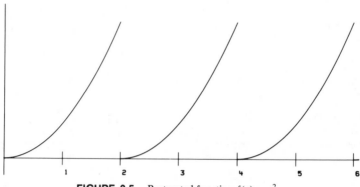

FIGURE 2.5. Protracted function $f(t) = t^2$.

phenomenon or *ringing* about the discontinuity. To mathematically appreciate this phenomenon, we require a rather sophisticated knowledge of integration theory beyond the scope of an applications text. However, here we heuristically note that it is caused because we are attempting to force a series of uniformly continuous functions of Equation (2.1) to converge to a discontinuous function. This is contrary to the basic truth that a series of uniformly continuous functions must converge to a function that is also uniformly continuous.

The fact that the functions (or protracted functions) are periodic allows us to shift the interval of integration by an arbitrary amount and not affect the value of the integral. To show this we first note:

$$\int_{-T/2+\delta}^{-T/2} f(t)\, dt = \int_{-T/2+\delta}^{-T/2} f(t+T)\, dt = \int_{T/2+\delta}^{T/2} f(\tau)\, d\tau = -\int_{T/2}^{T/2+\delta} f(\tau)\, d\tau.$$

In this equation we first made use of the fact that $f(t+T) = f(t)$ and then made the change of variable $\tau = t + T$ and changed the limits appropriately.

Now we write

$$\int_{-T/2+\delta}^{T/2+\delta} f(t)\, dt = \int_{-T/2+\delta}^{-T/2} f(t)\, dt + \int_{T/2}^{T/2+\delta} f(t)\, dt + \int_{-T/2}^{T/2} f(t)\, dt$$

$$= \int_{-T/2}^{T/2} f(t)\, dt.$$

Although not specifically pointed out, Equations (2.3) and (2.4) implied that the periodic function $f(t)$ was centered about $t = 0$ and that we performed the required integration between the limits of $-T/2$ and $T/2$. This previous result relaxes that condition. That is, $t = 0$ need not be the center of the period of $f(t)$. Furthermore, the limits of integration notation ($-T/2$ to $T/2$) now mean that we perform the required integration from the left end point to the right end point of the periodic interval and not necessarily from the specific values of $-T/2$ to $T/2$. The next two examples should clarify these remarks.

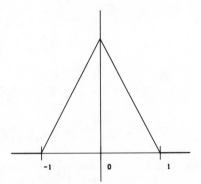

FIGURE 2.6. Example problem.

Let us first calculate the Fourier series of the function described by

(2.8)
$$f(t) = \begin{cases} 1 + t, & -1 \leqslant t \leqslant 0, \\ 1 - t, & 0 < t \leqslant 1, \\ 0, & \text{otherwise.} \end{cases}$$

This function is shown in Figure 2.6.

We proceed to generate the protracted function. It is somewhat arbitrary as to how we define the interval over which the function is to be integrated. In this example, let us consider $t = 0$ to be the center of the interval and consider the function from -1 to 1. Obviously, for this choice of an interval we have the period $T = 2$ and the frequency [Equation (2.2)], given by $\omega_n = n/T = n/2$. We first use Equation (2.3) to determine the pure cosine frequencies.

$$A_n = \frac{2}{T} \int_{-T/2}^{T/2} f(t) \cos \frac{2\pi nt}{T} \, dt, \qquad n = 1, 2, \ldots,$$

$$A_n = \frac{2}{2} \int_{-1}^{1} f(t) \cos \frac{2\pi nt}{2} \, dt,$$

$$= \int_{-1}^{0} (1 + t) \cos(\pi nt) \, dt + \int_{0}^{1} (1 - t) \cos(\pi nt) \, dt.$$

$$A_n = \frac{2 - 2 \cos \pi n}{(\pi n)^2}.$$

$$A_0 = \frac{1}{T} \int_{-T/2}^{T/2} f(t) \, dt = \frac{1}{2} \int_{-1}^{0} (1 + t) \, dt + \frac{1}{2} \int_{0}^{1} (1 - t) \, dt = \frac{1}{2}.$$

We point out here that the preceding formula could be reduced to the compact

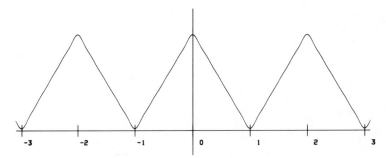

FIGURE 2.7. Fourier series approximation.

form

$$A_n = \begin{cases} 0, & n \text{ even,} \\ 4/(\pi n)^2, & n \text{ odd.} \end{cases}$$

However, for reasons that we explain later, we prefer to leave it in the general form. We now use Equation (2.4) to determine the pure sine frequencies.

$$B_n = \frac{2}{T}\int_{-T/2}^{T/2} f(t)\sin\frac{2\pi nt}{T}\,dt, \qquad n = 1, 2, \dots.$$

$$= \int_{-1}^{0}(1 + t)\sin(\pi nt)\,dt + \int_{0}^{1}(1 - t)\sin(\pi nt)\,dt = 0.$$

Substituting these expressions into Equation (2.1) we obtain

$$f(t) = \frac{1}{2} + \sum_{n=1}^{\infty}\left(\frac{2 - 2\cos\pi n}{(\pi n)^2}\right)\cos\pi nt.$$

The sum of the first 11 terms of this series approximation ($n = 0, 1, 2, \dots, 10$) is shown in Figure 2.7 over the interval from -3.1 to 3.1.

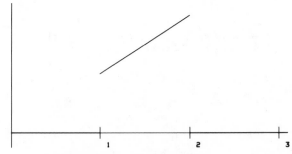

FIGURE 2.8. Another example problem.

As another example, let us determine the frequency content of the function $f(t) = t$ on the closed interval [1, 2]. This function is shown in Figure 2.8. For this function we have $T = 1$, $-T/2 = 1$, and $T/2 = 2$. Obviously, the protracted function is discontinuous and we should expect to see Gibbs phenomenon. Using Equation (2.3) we obtain the pure cosine terms as follows:

$$A_n = \frac{2}{T} \int_{-T/2}^{T/2} f(t)\cos\frac{2\pi nt}{T} \, dt = 2 \int_1^2 t\cos(2\pi nt) \, dt.$$

$$A_n = \left(\frac{\cos 4\pi n - \cos 2\pi n}{2(\pi n)^2} \right) + \left(\frac{2\sin 4\pi n - \sin 2\pi n}{\pi n} \right).$$

$$A_0 = \frac{1}{T} \int_{-T/2}^{T/2} f(t) \, dt = \int_1^2 t \, dt = 1.5.$$

Similarly, using Equation (2.4) for the pure sine terms we find

$$B_n = \frac{2}{T} \int_{-T/2}^{T/2} f(t)\sin\frac{2\pi nt}{T} \, dt = 2 \int_1^2 t\sin(2\pi nt) \, dt.$$

$$B_n = \left(\frac{\sin 4\pi n - \sin 2\pi n}{2(\pi n)^2} \right) + \left(\frac{\cos 2\pi n - 2\cos 4\pi n}{\pi n} \right).$$

Shown in Figure 2.9 is the sum of the first 11 terms ($n = 0, 1, \ldots, 10$) of the Fourier series approximation. Note the Gibbs phenomenon or ringing about the points of discontinuity.

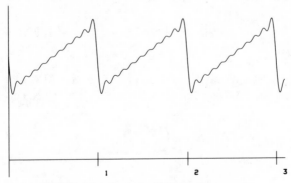

FIGURE 2.9. Example of Gibb's phenomenon.

COMPLEX FORM OF FOURIER SERIES

In Chapter 1 we were able to use a form of Euler's equations [Equations (1.22) and (1.23)] to write the general frequency content of a function in its complex form as per Equation (1.29). When we apply these equations to the Fourier series frequency content we obtain

(2.9)
$$f(t) = \sum_{n=-\infty}^{\infty} C_n e^{2\pi i n t/T},$$

where

$$C_n = \frac{A_n + iB_n}{2}, \qquad n < 0.$$

$$C_0 = A_0.$$

(2.10)

$$C_n = \frac{A_n - iB_n}{2}, \qquad n > 0.$$

$$\omega_n = \frac{n}{T}, \qquad n = \ldots -2, -1, 0, 1, 2, \ldots.$$

If we compare these equations to Equations (1.29), we see that they are quite similar. Other than the infinite range of the index n, the only real difference is the C_0 term. The reason is that frequency content as discussed in Chapter 1 did not have a specific formula for ω_n, whereas for Fourier series frequency content we do (namely, $\omega_n = n/T$). Thus the C_0 term corresponds to a 0 frequency or constant term. Equation (2.10) gives us the complex Fourier series coefficients C_k in terms of the pure cosine and pure sine Fourier coefficients. We now derive a more direct expression for the C_k terms. To do this we first require the following complex orthogonality relation:

(2.11)
$$\int_{-T/2}^{T/2} e^{2\pi i n t/T} e^{-2\pi i m t/T}\, dt = \begin{cases} 0, & n \neq m \\ T, & n = m. \end{cases}$$

We begin by first multiplying both sides of Equation (2.9) by $e^{-2\pi i m t/T}$ and then integrating the resulting equation from $-T/2$ to $T/2$. This results in:

$$\int_{-T/2}^{T/2} f(t) e^{-2\pi i m t/T} dt = \int_{-T/2}^{T/2} \sum_{n=-\infty}^{\infty} C_n e^{2\pi i n t/T} e^{-2\pi i m t/T} dt,$$

$$m = \ldots -2, -1, 0, 1, 2, \ldots.$$

Now using the previous orthogonality relation [Equation (2.11)], we see

(2.12) $\qquad C_n = \dfrac{1}{T}\displaystyle\int_{-T/2}^{T/2} f(t)e^{-2\pi i n t/T}dt, \qquad n = \ldots, -1, 0, 1, \ldots.$

Thus to calculate the complex Fourier series of a function we can proceed in one of two ways. We can first calculate the rectangular form (Equations 2.1–2.4) and then use Equations (2.10) to obtain the C_n terms. The other approach is to calculate these terms directly from Equation (2.12). This is best illustrated by an example. Let us calculate the complex Fourier series frequency content of the function given by Equation (2.8). From a previous example we determined

$$A_n = \frac{2 - 2\cos \pi n}{(\pi n)^2}, \qquad n = 1, 2, \ldots.$$

$$A_0 = \frac{1}{2}$$

$$B_n = 0.$$

Thus Equation (2.10) implies,

$$C_n = \frac{1 - \cos \pi n}{(\pi n)^2}, \qquad n > 0,$$

$$C_0 = \frac{1}{2},$$

$$C_n = \frac{1 - \cos \pi n}{(\pi n)^2}, \qquad n < 0.$$

We now calculate these complex coefficients directly using Equation (2.12).

$$C_n = \frac{1}{T}\int_{-T/2}^{T/2} f(t)e^{-2\pi i n t/T}dt,$$

$$C_n = \frac{1}{2}\int_{-1}^{1} f(t)e^{-i\pi n t}dt,$$

$$C_n = \frac{1}{2}\int_{-1}^{0}(1 + t)e^{-i\pi n t}dt + \frac{1}{2}\int_{0}^{1}(1 - t)e^{-i\pi n t}dt,$$

$$C_n = \frac{1}{2}\left(\frac{e^{i\pi n} + e^{-i\pi n} - 2}{(i\pi n)^2}\right).$$

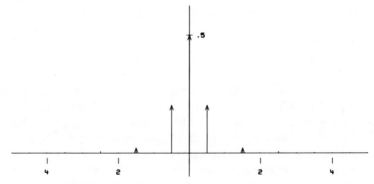

FIGURE 2.10.   Frequency domain plot of equation (2.8).

Thus

$$C_n = \frac{1 - \cos \pi n}{(\pi n)^2}, \qquad n = \ldots, -2, -1, 0, 1, 2, \ldots .$$

We note that this equation is valid for all values of n, including $n = 0$. Therefore, we can obtain the C_0 term from this equation (use L'Hôpital's limit rule twice), or we can use the equation

$$C_0 = \frac{1}{T} \int_{-T/2}^{T/2} f(t)\, dt = \frac{1}{2} \int_{-1}^{0} (1 + t)\, dt + \frac{1}{2} \int_{0}^{1} (1 - t)\, dt = \frac{1}{2}.$$

Shown in Figure 2.10 is the frequency domain plot for this function from $n = -5$ to 5. *Note*: The C_n terms are pure real and, therefore, we only require one frequency domain plot (the imaginary plot is everywhere zero).

PROPERTIES OF THE FOURIER SERIES

The main concern to most users of Fourier series is the calculation of the constants A_n and B_n or C_n. The integral expressions for these constants are sometimes quite involved and evaluation of the limits are difficult. Thus if we can in any way avoid the evaluation of the integrals, it is to our advantage. We now discuss some of the useful properties of the Fourier series that usually simplify the calculation of the constants. These properties are presented as theorems, the reason being that this approach offers the most compact and easily referenced way to catalogue our results. With few exceptions a formal proof is not given but instead only a heuristic discussion of what the theorem means and how it can be used is presented.

Theorem 2.1 (Linearity). If both functions $f(t)$ and $g(t)$ have Fourier series frequency content given by the sets $\{A_n, B_n, w_n\}$ and $\{C_n, D_n, w_n\}$, respectively, then the function $h(t) = af(t) + bg(t)$ has Fourier series frequency content given by the set $\{aA_n + bC_n, aB_n + bD_n, w_n\}$.

This theorem permits us to calculate the Fourier series of a function that is a linear combination of two other functions that possess Fourier series. We need not go through all the work of evaluating the integrals of Equations (2.3) and (2.4) but instead we perform a simple addition of two sets. In other words, if the function $f(t)$ has Fourier series coefficients given by A_k and B_k and the function $g(t)$ has coefficients given by C_k and D_k, then the function $af(t) + bg(t)$ has coefficients given by $aA_k + bC_k$ and $aB_k + bD_k$. Similar results are also true for the exponential form:

Corollary 2.1. If both $f(t)$ and $g(t)$ have complex Fourier series coefficients given by F_k and G_k, respectively, then the function $h(t) = af(t) + bg(t)$ has Fourier series coefficients given by $aF_k + bG_k$.

Theorem 2.2 (First Shifting Theorem). If $f(t)$ has complex frequency content given by the set $\{F_k, \omega_k\}$ and a is a constant, then the complex frequency content of the function $f(t - a)$ is given by the set

$$\left\{ F_k e^{-2\pi ika/T}, \omega_k \right\}.$$

When converted from polar to rectangular form Theorem 2.2 results in the following.

Corollary 2.2. If $f(t)$ has frequency content given by the set $\{A_k, B_k, \omega_k\}$ and a is a constant, then the frequency content of the function $f(t - a)$ is given by the set:

$$\left\{ A_k \cos\frac{2\pi ka}{T} - B_k \sin\frac{2\pi ka}{T}, A_k \sin\frac{2\pi ka}{T} + B_k \cos\frac{2\pi ka}{T}, \omega_k \right\}.$$

This first shifting theorem provides us with a method of obtaining the Fourier series (rectangular or complex form) of a function $f(t - a)$ from the Fourier series coefficients of $f(t)$. That is, if we know the Fourier series expansion of $f(t)$, then we also know the Fourier series expansion of the function $f(t - a)$ and, in fact, the coefficients are given by the relatively simple expressions presented in Theorem 2.2 and Corollary 2.2.

FOURIER SERIES OF COMPLEX FUNCTIONS

Thus far we have implicitly considered the function f to be real valued. We now explicitly consider the situation when f is a complex function. That is, when f can be written as

$$(2.13) \qquad f(t) = f_1(t) + i f_2(t),$$

where $f_1(t)$ and $f_2(t)$ are both real functions. Furthermore, we insist that both f_1 and f_2 are periodic with period T and both possess Fourier series frequency content. In this case it is not difficult to show that f also has frequency content. We proceed as follows:

$$f_1(t) = \sum_{k=-\infty}^{\infty} A_{1,k} \cos \frac{2\pi k t}{T} + B_{1,k} \sin \frac{2\pi k t}{T},$$

$$f_2(t) = \sum_{k=-\infty}^{\infty} A_{2,k} \cos \frac{2\pi k t}{T} + B_{2,k} \sin \frac{2\pi k t}{T},$$

where

$$A_{1,k} = \frac{2}{T} \int_{-T/2}^{T/2} f_1(t) \cos \frac{2\pi k t}{T} \, dt,$$

$$B_{1,k} = \frac{2}{T} \int_{-T/2}^{T/2} f_1(t) \sin \frac{2\pi k t}{T} \, dt,$$

$$A_{2,k} = \frac{2}{T} \int_{-T/2}^{T/2} f_2(t) \cos \frac{2\pi k t}{T} \, dt,$$

$$B_{2,k} = \frac{2}{T} \int_{-T/2}^{T/2} f_2(t) \sin \frac{2\pi k t}{T} \, dt.$$

We now wish to find constants A_k and B_k such that

$$f(t) = \sum_{k=-\infty}^{\infty} A_k \cos \frac{2\pi k t}{T} + B_k \sin \frac{2\pi k t}{T} \, dt.$$

To calculate the pure cosine terms we simply use Equation (2.3) (as if f were a real function) and then take advantage of the linearity Theorem 2.1:

$$A_k = \frac{2}{T} \int_{-T/2}^{T/2} f(t) \cos \frac{2\pi k t}{T} \, dt,$$

$$= \frac{2}{T} \int_{-T/2}^{T/2} f_1(t) \cos \frac{2\pi k t}{T} \, dt + i \frac{2}{T} \int_{-T/2}^{T/2} f_2(t) \cos \frac{2\pi k t}{T} \, dt.$$

Thus

(2.14)
$$A_k = A_{1,k} + iA_{2,k}.$$

In a similar manner we can show that the pure sine terms B_k are given as

(2.15)
$$B_k = B_{1,k} + iB_{2,k}.$$

We see, therefore, that the coefficients A_k and B_k are calculated in exactly the same way as if f were a real function. When f, in fact, is a complex function, these coefficients turn out to be complex numbers as given by Equations (2.14 and 2.15). In effect we obtain two real Fourier series if f is complex. When f is a complex function, the complex form of the Fourier series [Equations (2.9) and (2.10)] turns out to be more convenient. It can easily be shown that if f is complex, as per Equation (2.13), then it can be represented as

$$f(t) = \sum_{k=-\infty}^{\infty} C_k e^{2\pi i k t/T},$$

with coefficients C_k given by

(2.16)
$$C_k = C_{1,k} + iC_{2,k},$$

where

$$C_{1,k} = \frac{1}{T} \int_{-T/2}^{T/2} f_1(t) e^{-2\pi i k t/T},$$

and

$$C_{2,k} = \frac{1}{T} \int_{-T/2}^{T/2} f_2(t) e^{-2\pi i k t/T}.$$

We have just illustrated that even when f is a complex function we can still obtain the Fourier series coefficients by using Equations (2.3) and (2.4) or, in the complex form, Equation (2.12). Therefore, from here on, we consider all our functions to be complex unless specifically stated otherwise. We point out that Theorems 2.1 and 2.2 (along with their corollaries) are also valid for complex functions.

ADDITIONAL PROPERTIES OF THE FOURIER SERIES

The ability of considering complex functions yields many other useful results. The first such one that we present is another shifting theorem.

Theorem 2.3 (Second Shifting Theorem).   If the function $f(t)$ has complex frequency content given by the set $\{C_k, \omega_k\}$, then the complex frequency content of the function

$$f(t)\dot{e}^{2\pi iat/T}, \quad \text{where } a \text{ is a constant,}$$

is given by the set $\{C_{k-a}, \omega_k\}$.

Thus we see if the complex Fourier series coefficients of a function $f(t)$ are given by C_k, then we obtain the complex coefficients for the function,

$$f(t)e^{2\pi iat/T}$$

by simply substituting $k - a$ for k in the expression for the C_k terms. For example, in a previous illustration we determined the complex Fourier series coefficients for the function in Equation (2.8) to be given by

$$C_k = \frac{1 - \cos \pi k}{(\pi k)^2} \quad \text{for all } k.$$

Thus the complex Fourier series coefficients of the function given by

$$f(t) = \begin{cases} (1 + t)e^{i\pi t/2}, & -1 \leqslant t \leqslant 0, \\ (1 - t)e^{i\pi t/2}, & 0 < t \leqslant 1, \\ 0, & \text{otherwise} \end{cases}$$

are determined by the formula,

$$D_k = \frac{1 - \cos(\pi k - 0.5\pi)}{\pi^2(k^2 - k + 0.25)}.$$

A brief word of caution: Some care must be exercised when applying the results of a theorem of this type. These substitutions must be made into a general formula for the coefficients and not into a simplified expression. For example, in the previous illustration we were able to substitute $k - a$ for k into the expression for C_k because this was a general formula; that is, it was not reduced to the simpler form

$$C_k = \begin{cases} 0, & k \text{ even,} \\ 2/(\pi k)^2, & k \text{ odd.} \end{cases}$$

Most authors choose to "simplify" their expressions whenever possible. Although this leads to neater results, it also introduces the danger of misapplication of theorems, such as those presented in this chapter.

The next two theorems are the rectangular form versions of the second shifting theorem.

Theorem 2.4. If the function $f(t)$ has frequency content given by the set $\{A_k, B_k, \omega_k\}$, then the frequency content of the function

$$f(t)e^{2\pi i a t/T}, \quad \text{where } a \text{ is a constant,}$$

is given by the set $\{\mathcal{Q}_k, \mathcal{B}_k, \omega_k\}$, where,

$$\mathcal{Q}_k = \frac{1}{2}\left[A_{k+a} + A_{k-a} + i(B_{k+a} - B_{k-a})\right]$$

$$\mathcal{B}_k = \frac{1}{2}\left[B_{k+a} + B_{k-a} + i(A_{k-a} - A_{k+a})\right].$$

Theorem 2.5. If f is a real valued function with frequency content given by the set $\{A_k, B_k, \omega_k\}$ and a is a constant, then

(i) $f(t)\cos(2\pi a t/T)$ has frequency content given by the set

$$\left\{\frac{A_{k+a} + A_{k-a}}{2}, \frac{B_{k+a} + B_{k-a}}{2}, \omega_k\right\};$$

(ii) $f(t)\sin(2\pi a t/T)$ has frequency content given by the set

$$\left\{\frac{B_{k+a} - B_{k-a}}{2}, \frac{A_{k-a} - A_{k+a}}{2}, \omega_k\right\}.$$

FOURIER SERIES OF ODD AND EVEN FUNCTIONS

Odd and even functions occur so often and offer such simplification that they are worthy of our special attention. A function f is called even if and only if $f(-t) = f(t)$, and similarly called odd if and only if $f(-t) = -f(t)$. Note that by their very nature, odd and even functions are centered (about 0) in the middle of their period. Consequently, the limits of integration in Equations (2.3) and (2.4) do in fact refer to the *values* $-T/2$ and $T/2$.

We first consider the Fourier series expansion of an even function. Splitting the integral in Equation (2.3) we have

$$A_k = \frac{2}{T}\int_{-T/2}^{0} f(t)\cos\frac{2\pi kt}{T}\,dt + \frac{2}{T}\int_{0}^{T/2} f(t)\cos\frac{2\pi kt}{T}\,dt.$$

If we now substitute $-t$ for t and $-dt$ for dt in the first integral, we obtain

$$A_k = \frac{-2}{T} \int_{T/2}^{0} f(-t) \cos \frac{-2\pi kt}{T} dt + \frac{2}{T} \int_{0}^{T/2} f(t) \cos \frac{2\pi kt}{T} dt.$$

Now, inverting the limits on the first integral (which changes the sign of the integral) and noting that $f(-t) = f(t)$ and $\cos(-2\pi kt/T) = \cos(2\pi kt/T)$ we find

$$A_k = \frac{2}{T} \int_{0}^{T/2} f(t) \cos \frac{2\pi kt}{T} dt + \frac{2}{T} \int_{0}^{T/2} f(t) \cos \frac{2\pi kt}{T} dt,$$

$$A_k = \frac{4}{T} \int_{0}^{T/2} f(t) \cos \frac{2\pi kt}{T} dt.$$

Now, for the B_k coefficients,

$$B_k = \frac{2}{T} \int_{-T/2}^{0} f(t) \sin \frac{2\pi kt}{T} dt + \frac{2}{T} \int_{0}^{T/2} f(t) \sin \frac{2\pi kt}{T} dt.$$

Again, substitution of $-t$ for t and $-dt$ for dt in the first integral of the previous equation yields

$$B_k = \frac{-2}{T} \int_{T/2}^{0} f(-t) \sin \frac{-2\pi kt}{T} dt + \frac{2}{T} \int_{0}^{T/2} f(t) \sin \frac{2\pi kt}{T} dt.$$

We invert the limits on the first integral and recall $f(t) = f(-t)$ and note that $\sin(-2\pi kt/T) = -\sin(2\pi kt/T)$. This results in

$$B_k = \frac{-2}{T} \int_{0}^{T/2} f(t) \sin \frac{2\pi kt}{T} dt + \frac{2}{T} \int_{0}^{T/2} f(t) \sin \frac{2\pi kt}{T} dt,$$

or

$$B_k = 0.$$

The evaluation of the Fourier series coefficients (in rectangular form) is significantly simplified for an even function. All the B_k coefficients vanish and the A_k terms are now calculated by performing the integration from 0 to $T/2$ (rather than $-T/2$ to $T/2$) and then doubling the value of the results. Analogous reasoning shows that for an odd function

$$A_k = 0,$$

$$B_k = \frac{4}{T} \int_{0}^{T/2} f(t) \sin \frac{2\pi kt}{T} dt.$$

We now summarize these results in the following two theorems.

Theorem 2.6 (Even Function). If $f(t)$ is an even periodic function, then its Fourier series coefficients are given by

$$A_0 = \frac{2}{T} \int_0^{T/2} f(t)\, dt,$$

$$A_n = \frac{4}{T} \int_0^{T/2} f(t)\cos\frac{2\pi nt}{T}\, dt, \qquad n = 1, 2, \ldots$$

$$B_n = 0 \quad \text{for all } n.$$

Theorem 2.7 (Odd Function). If $f(t)$ is an odd periodic function then its Fourier series coefficients are given by

$$B_0 = 0,$$

$$B_n = \frac{4}{T} \int_0^{T/2} f(t)\sin\frac{2\pi nt}{T}\, dt, \qquad n = 1, 2, \ldots$$

$$A_n = 0 \quad \text{for all } n.$$

The fact that we require only a portion of the work in evaluating the Fourier series coefficients of odd and even functions makes them quite useful. Let us illustrate this with an example in which we determine the A_k and B_k terms of the function given by

$$f(t) = \begin{cases} \sin 2\pi t, & 0 \leqslant t \leqslant \tfrac{1}{2}, \\ 0, & \text{otherwise.} \end{cases}$$

The interval of interest is $[-\tfrac{1}{2}, \tfrac{1}{2}]$. The protracted version of this function is shown in Figure 2.11. As can be seen, this function is neither odd nor even.

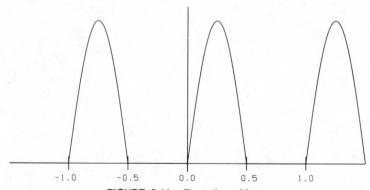

FIGURE 2.11. Example problem.

Therefore, to take advantage of the previous theorems, we must first perform some mathematical manipulations. We begin by considering the function

$$g(t) = \begin{cases} \cos 2\pi t, & -\frac{1}{4} \leqslant t \leqslant \frac{1}{4}, \\ 0, & \text{otherwise.} \end{cases}$$

The interval of interest is $[-\frac{1}{2}, \frac{1}{2}]$. Obviously, this is the same function that we previously considered shifted to the left by an amount $\frac{1}{4}$. The protracted version of this function is shown in Figure 2.12. Inasmuch as this is an even function, we need only consider the pure cosine $\mathcal{Q}_k$ terms as per Theorem 2.6. Thus

$$\mathcal{Q}_k = \frac{4}{T} \int_0^{T/2} g(t) \cos \frac{2\pi kt}{T} \, dt = 4 \int_0^{1/4} \cos(2\pi t) \cos(2\pi kt) \, dt.$$

Now going through the required moves we obtain

$$\mathcal{Q}_0 = \frac{1}{\pi},$$

$$\mathcal{Q}_k = \frac{\sin[(1+k)\pi/2]}{(1+k)\pi} + \frac{\sin[(1-k)\pi/2]}{(1-k)\pi}.$$

We now use the first shifting theorem (Corollary 2.2) to obtain the coefficients of $f(t) = g(t-a)$. First we note that $\mathcal{Q}_0 = 1/\pi$ is not a function of k and thus $A_0 = \mathcal{Q}_0 = 1/\pi$. Now for values of $k \neq 1$ we have

$$A_k = \left\{ \frac{\sin[(1+k)\pi/2]}{(1+k)\pi} + \frac{\sin[(1-k)\pi/2]}{(1-k)\pi} \right\} \cos \frac{\pi k}{2}.$$

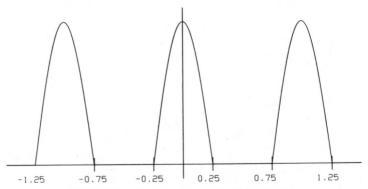

FIGURE 2.12. Example problem of a shifted function.

This can be simplified by using a variation of the formula $\sin(\alpha + \beta) = \sin\alpha\cos\beta + \cos\alpha\sin\beta$:

$$A_k = \frac{\sin[(1 + k)\pi/2]\cos(\pi k/2)}{(1 + k)\pi} + \frac{\sin[(1 - k)\pi/2]\cos(\pi k/2)}{(1 - k)\pi}$$

$$A_k = \frac{1}{2}\left\{\frac{\sin[(1 + 2k)\pi/2] + \sin(\pi/2)}{(1 + k)\pi}\right\}$$

$$+ \frac{1}{2}\left\{\frac{\sin(\pi/2) + \sin[(1 - 2k)\pi/2]}{(1 - k)\pi}\right\}.$$

If we note that $\sin(\pi/2) = 1$, $\sin[(1 + 2k)\pi/2] = -\cos(1 + k)\pi$, and $\sin[(1 - 2k)\pi/2] = -\cos(1 - k)\pi$, then we obtain

$$A_k = \frac{1 - \cos(1 + k)\pi}{2\pi(1 + k)} + \frac{1 - \cos(1 - k)\pi}{2\pi(1 - k)}.$$

As for the pure sine B_k terms, Corollary 2.2 implies

$$B_k = \left\{\frac{\sin[(1 + k)\pi/2]}{(1 + k)\pi} + \frac{\sin[(1 - k)\pi/2]}{(1 - k)\pi}\right\}\sin\frac{\pi k}{2}.$$

As can be seen when k is even, the term $\sin(k\pi/2)$ vanishes and, similarly, when k is odd ($k \neq 1$), the $\sin[(1 + k)\pi/2]$ term vanishes as well as the $\sin[(1 - k)\pi/2]$ term. The only situation we need explore is

$$\left\{\frac{\sin[(1 - k)\pi/2]}{(1 - k)\pi}\right\}\sin\frac{k\pi}{2} \quad \text{for } k = 1.$$

When $k = 1$ we have the indeterminate form $\frac{0}{0}$ and thus we must apply L'Hôpital's rule for the limiting case of $k = 1$. Doing this we find that $B_1 = 1/2$.

THE FUNCTION sin 2πat

In this section we discuss the frequency content of the function $\sin 2\pi at$, not only because it is useful but also because this example helps illustrate several interesting features of Fourier series frequency content. For the sake of discussion we fix the period T to be 1 and first consider the case when a is equal to an integer m. Such a function is shown in Figure 2.13 and the

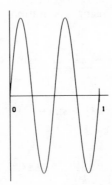

FIGURE 2.13. $f(t) = \sin 2\pi m t$.

protracted version of the function is shown in Figure 2.14. We first note that
since $\sin 2\pi a t$ is an odd function we only have B_k terms. Thus

$$B_k = 4\int_0^{1/2}\sin(2\pi m t)\sin(2\pi k t)\ dt.$$

The orthogonality equation (2.6) implies that this integral has a value only
when $k = m$ and thus, as was obvious from the start, this function has only
one frequency (monochromatic function). We now look at the more interesting
case when a is not an integer. Shown in Figure 2.15 is the situation when $a > 1$
and in Figure 2.16, the protracted version. The first fact that becomes apparent
about this function is that it contains discontinuities. We now wish to find the
Fourier series coefficients of this function. We see that this function is neither
even nor odd; therefore, we must calculate both the A_k and B_k coefficients. We

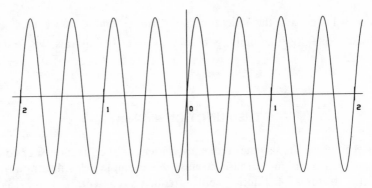

FIGURE 2.14.   Protracted version of $\sin 2\pi m t$.

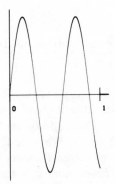

FIGURE 2.15. $f(t) = \sin 2\pi at$.

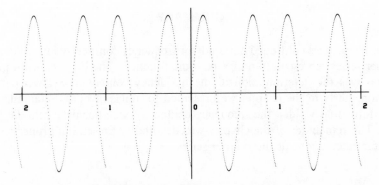

FIGURE 2.16. Protracted version of $\sin 2\pi at$.

proceed as follows:

$$A_k = 2\int_0^1 \sin(2\pi at)\cos(2\pi kt)\, dt,$$

$$A_k = \int_0^1 \sin[2\pi(a+k)t]\, dt + \int_0^1 \sin[2\pi(a-k)t]\, dt.$$

$$A_k = \frac{1 - \cos[2\pi(a+k)t]}{2\pi(a+k)} + \frac{1 - \cos[2\pi(a-k)t]}{2\pi(a-k)}.$$

$$A_0 = \int_0^1 \sin(2\pi at)\, dt = \frac{1 - \cos 2\pi a}{2\pi a}.$$

Similarly, we find

$$B_k = \frac{\sin[2\pi(a - k)]}{2\pi(a - k)} - \frac{\sin[2\pi(a + k)]}{2\pi(a + k)}$$

We recall that when a was an integer we had only one frequency. However, in this case we see that we, indeed, have many frequencies required to construct the function $\sin 2\pi at$. These additional frequencies are required to fit the sharp discontinuities at 0 and 1. We note that the closer that a is to an integer, the smaller the values of the coefficients become. Thus the smaller the jump at the discontinuity, the fewer number of terms are required to accurately "track" the function.

SUMMARY

In this chapter we continued our study of frequency content by introducing the Fourier series representation of a periodic function. The basic concern in this chapter was the determination of the frequency content of a given periodic function. Early in the chapter we presented formulas in both rectangular and polar form that yielded the frequency content of well-behaved periodic functions. The remainder of the chapter was devoted to presenting properties and example calculations of the Fourier series.

BIBLIOGRAPHY

Apostal, T. M., *Mathematical Analysis*, 1st ed. Addison-Wesley, Reading, Mass., 1957.

Raven, F. H., *Mathematics of Engineering Systems*, McGraw-Hill, New York, 1966.

Churchill, R. V., *Fourier Series and Boundary Value Problems*, McGraw-Hill, New York, 1963.

Ritt, R. V., *Fourier Series*, McGraw-Hill, New York, 1970.

Tolstov, G. P., *Fourier Series*, Prentice-Hall, Englewood Cliffs, N.J., 1962.

Worsnop, B. L., *An Introduction to Fourier Analysis*, Wiley, New York, 1961.

CHAPTER

3

THE FOURIER
TRANSFORM

In the previous chapter we discussed the subject of frequency content of periodic functions. The focus of that study was the Fourier series representation of (periodic) functions. In this chapter we consider the frequency content of functions that are not required to be periodic. The focus of this study is the Fourier transform representation of a function. Given a function f, we define the *Fourier transform pair* as

(3.1)
$$F(w) = \int_{-\infty}^{\infty} f(t) e^{-2\pi i w t} \, dt,$$

(3.2)
$$f(t) = \int_{-\infty}^{\infty} F(w) e^{2\pi i w t} \, dw. \quad \leftarrow \text{Inverse}$$

Equation (3.1) is called the (direct) Fourier transform and Equation (3.2) is known as the inverse Fourier transform. We say $F(w)$ is the Fourier transform of $f(t)$ and that $f(t)$ is the inverse Fourier transform of $F(w)$. Notationally, we write

$$F(w) = \mathcal{F}[f(t)],$$

$$f(t) = \mathcal{F}^{-1}[F(w)].$$

The existence of these integrals is an entire topic by itself. Unfortunately, it is beyond the scope of an applications text such as this and, consequently, we make the very important assumption that the integrals of both these equations exist. We also assume that each equation implies the other. That is to say, if we obtain $F(w)$ from Equation (3.1), then Equation (3.2) will uniquely return $f(t)$. Similarly, if we obtain $f(t)$ from Equation (3.2), then Equation (3.1) will return $F(w)$ uniquely.

We could be content with just defining these equations as we have done, and then proceed to derive useful properties and present examples. However, inasmuch as we wish to preserve the concept of frequency content in this chapter, we show (in a rather loose fashion) that these equations follow directly from the Fourier series as the period T approaches infinity. Our starting point is the complex form of the Fourier series:

$$(3.3) \qquad f(t) = \sum_{k=-\infty}^{\infty} C_k e^{2\pi i k t / T},$$

$$(3.4) \qquad C_k = \frac{1}{T} \int_{-T/2}^{T/2} f(t) e^{-2\pi i k t / T} \, dt.$$

We recall that the constants C_k are a measure of the "amount" of the discrete frequencies ($w_k = k/T$) that are combined to represent the periodic function $f(t)$. We also note that although we have an infinite number of these discrete frequencies, they are all a multiple of a basic frequency $w_0 = 1/T$. As the period T increases, this basic frequency decreases and, therefore, the discrete frequencies become closer together until in the limit ($T \to \infty$) they equal a continuous spectrum. That is,

$$\lim_{T \to \infty} w_k = w, \qquad k = \cdots - 1, 0, 1, \ldots .$$

This limiting procedure is schematically illustrated in Figure 3.1. We also note that

$$\Delta w_k = w_{k+1} - w_k = \frac{(k+1)}{T} - \frac{k}{T} = \frac{1}{T}.$$

Now let us multiply both sides of Equation (3.4) by T to obtain

$$TC_k = \int_{-T/2}^{T/2} f(t) e^{-2\pi i k t / T} \, dt$$

and thus in the limit as $T \to \infty$ we have

$$F(w) = \lim_{T \to \infty} (TC_k) = \int_{-\infty}^{\infty} f(t) e^{-2\pi i w t} \, dt.$$

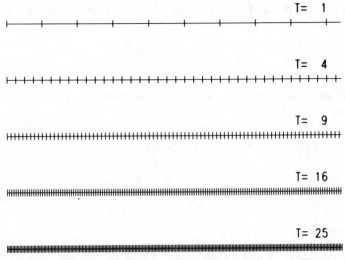

FIGURE 3.1. Discrete to continuous spectrum ($\Delta w_K = 1/T$).

This establishes Equation (3.1). In Equation (3.3) if we multiply and divide the right-hand side by T, we obtain

$$f(t) = \sum_{k=-\infty}^{\infty} TC_k e^{2\pi ikt/T} \frac{1}{T} = \sum_{k=-\infty}^{\infty} TC_k e^{2\pi ikt/T} \Delta w_k.$$

In the limit as $T \to \infty$ we have

$$TC_k = F(w),$$

$$w_k = \frac{k}{T} = w,$$

$$\Delta w_k = dw.$$

Also, the summation may be considered to approach an integral and thus we obtain Equation (3.2).

FOURIER TRANSFORM FREQUENCY CONTENT

We now give the Fourier transform a frequency content interpretation. To do this, we first rewrite Equation (3.2) in the form

$$(3.5) \qquad f(t) = \int_0^\infty \left[C(w)\cos 2\pi wt + S(w)\sin 2\pi wt \right] dw,$$

in which case we define the *Fourier transform frequency content* as the infinite set of 3-tuples

(3.6) $$\{C(w), S(w), w\}.$$

$C(w)$ is called the pure cosine frequency content of frequency w and $S(w)$ is called the pure sine frequency content of frequency w. We obtained Equation (3.5) by rewriting Equation (3.2) as

$$f(t) = \int_{-\infty}^{\infty} F(w)(\cos 2\pi wt + i \sin 2\pi wt)\, dw,$$

$$f(t) = \int_{0}^{\infty} \{[F(w) + F(-w)]\cos 2\pi wt + i[F(w) - F(-w)]\sin 2\pi wt\}\, dw.$$

Thus we see

(3.7) $$C(w) = F(w) + F(-w),$$

(3.8) $$S(w) = i[F(w) - F(-w)].$$

Proceeding, we can also show

$$C(w) = F(w) + F(-w),$$

$$C(w) = \int_{-\infty}^{\infty} f(t)e^{-2\pi iwt}\, dt + \int_{-\infty}^{\infty} f(t)e^{2\pi iwt}\, dt,$$

$$C(w) = 2\int_{-\infty}^{\infty} f(t)\left(\frac{e^{-2\pi iwt} + e^{2\pi iwt}}{2}\right)\, dt,$$

(3.9) $$C(w) = 2\int_{-\infty}^{\infty} f(t)\cos(2\pi wt)\, dt.$$

Similarly, we can show that

(3.10) $$S(w) = 2\int_{-\infty}^{\infty} f(t)\sin(2\pi wt)\, dt.$$

We have just illustrated that the pure sine and pure cosine frequency content terms $S(w)$ and $C(w)$ can be calculated in either of two ways. We can calculate them from the Fourier transform as per Equations (3.7) and (3.8), or directly by means of Equations (3.9) and (3.10).

NOTE: In Equations (3.9) and (3.10) when the lower limit of integration is zero, then $C(w)$ and $S(w)$ are called the Fourier cosine transform and Fourier sine transform, respectively.

If we compare the Fourier transform frequency content (Equation 3.6) to the Fourier series frequency content ($\{A_k, B_k, w_k\}$) as defined in Chapter 2, we note the similarities. The Fourier series frequency content is an infinite set consisting of a discrete frequency spectrum, whereas the Fourier transform frequency content is an infinite set made up of a continuous frequency spectrum. A plot of $F(w)$ versus w is called the frequency domain plot of f and we can readily see that it is completely analogous to the Fourier series frequency domain plot of C_k versus w_k.

We now present an example that illustrates both the calculation of the Fourier transform of a function and the similarity that it has to the Fourier series of that same function. To do this let us consider the pulse function $p_a(t)$ described by

$$p_a(t) = \begin{cases} 1, & |t| \leqslant a, \\ 0, & \text{otherwise.} \end{cases}$$

The Fourier transform of this function is given by

$$F(w) = \int_{-\infty}^{\infty} p_a(t) e^{-2\pi i w t} \, dt = \int_{-a}^{a} e^{-2\pi i w t} \, dt,$$

$$F(w) = \frac{e^{2\pi i w a} - e^{-2\pi i w a}}{2\pi i w},$$

(3.11) $$F(w) = 2a \frac{\sin 2\pi w a}{2\pi w a} = 2a \operatorname{sinc} 2\pi w a.$$

Thus we see that the frequency spectrum of this pulse function is given by the continuous real function $2a \operatorname{sinc} 2\pi w a$. The frequency domain plot is shown in Figure 3.2.

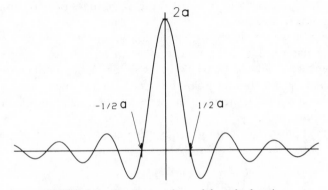

FIGURE 3.2. Fourier transform of the pulse function.

Let us compare this frequency spectrum to that which we obtain by using the Fourier series frequency content. To calculate the Fourier series of this pulse we must first generate the protracted, or periodic, version of this function. To do this we define the step function over the period T:

$$p_b(t) = \begin{cases} 1, & -bT \leqslant t \leqslant bT \quad (0 \leqslant b \leqslant \frac{1}{2}), \\ 0, & \text{otherwise.} \end{cases}$$

Using the complex form of the Fourier series [Equation (2.12)] we find

$$C_k = \frac{1}{T} \int_{-T/2}^{T/2} f(t) e^{-2\pi i k t/T} dt = \frac{1}{T} \int_{-bT}^{bT} e^{-2\pi i k t/T} dt,$$

or

(3.12) $$C_k = 2b \operatorname{sinc} 2\pi bk \quad \text{for all } k.$$

We first note that both the Fourier transform frequency spectrum and the Fourier series frequency spectrum have the same basic or overall shape (i.e., a sinc function). For a more illustrative comparison, let us choose a specific value for the pulse half-width, namely, $a = 1$. In this case the Fourier transform spectrum (Equation 3.11) becomes

$$F(w) = 2 \operatorname{sinc} 2\pi w.$$

When dealing with the Fourier series or protracted function we require that $bT = a = 1$ and $(0 \leqslant b \leqslant \frac{1}{2})$. We have quite a bit of freedom in choosing the period T and fraction b such that the preceding equation is true. However, just to be specific, let us choose $T = 4$ and $b = .25$. Equation (3.12) then becomes

$$C_k = \frac{1}{2} \operatorname{sinc} \frac{\pi k}{2}.$$

In Figure 3.3 we plot both $F(w)$ versus w and C_k versus $w_k = k/T$ together. For illustrative purposes we have scaled the magnitudes of $F(w)$ and C_k to be the same in this figure.

In this figure we clearly see the continuous nature of $F(w)$ contrasted with the discrete nature of C_k. We note that when an infinite number of these discrete frequencies are combined as per Equation (2.9), we obtain the periodic or protracted version of the pulse function. On the other hand, when we combine the continuous spectrum as per equation (3.2), we obtain only the basic pulse function.

As our next illustrative example let us calculate the Fourier transform of the function:

$$f(t) = \begin{cases} e^{-at}, & t \geqslant 0 (a > 0), \\ 0, & \text{otherwise,} \end{cases}$$

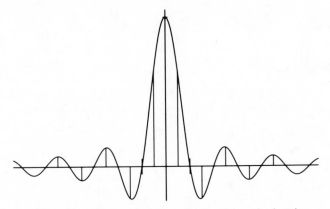

FIGURE 3.3. Fourier transform and series of the pulse function.

shown in Figure 3.4. Using Equation (3.1) we obtain

$$F(w) = \int_{-\infty}^{\infty} f(t)e^{-2\pi iwt}\, dt = \int_{0}^{\infty} e^{-at}e^{-2\pi iwt}\, dt$$

We now see the reason for requiring $a > 0$ in the definition of this function. Obviously, if $a < 0$, then e^{-at} would be unbounded and this integral would not exist. Evaluating this integral we obtain

$$F(w) = \frac{1}{a + 2\pi iw} = \frac{a - 2\pi iw}{a^2 + 4\pi^2 w^2}.$$

This function is obviously complex and is shown in Figure 3.5 with the solid line representing the shape of the real portion and the dashed line representing the shape of the imaginary part.

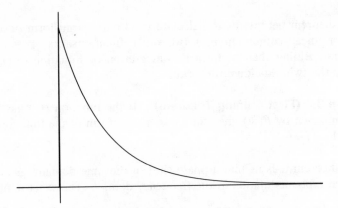

FIGURE 3.4. $f(t) = e^{-at}$.

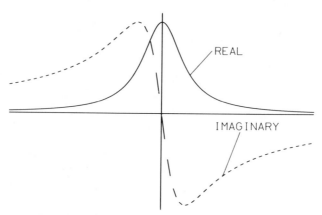

FIGURE 3.5. Fourier transform of e^{-at}.

PROPERTIES OF THE FOURIER TRANSFORM

Given a function f, the problem is to calculate its Fourier transform as per Equation (3.1). Often evaluation of the integral in this equation is very tedious and involved. Therefore, we present several theorems that discuss properties of the transform that can usually be used to simplify the calculations. Just as we did in Chapter 2 for the Fourier series, we present these results as theorems because this approach affords the most compact and easily referenced way to catalogue our results. With few exceptions, a formal proof is not given but instead a heuristic discussion of what the theorem means and how it can be used is presented.

Theorem 3.1 (Linearity). If both functions $f(t)$ and $g(t)$ have Fourier transforms given by $F(w)$ and $G(w)$, respectively, then the function $h(t) = af(t) + bg(t)$ has a Fourier transform given by $aF(w) + bG(w)$.

This theorem permits us to calculate the Fourier transform of a function that is a linear combination of two other functions that possess Fourier transforms. Rather than perform the integration of Equation (3.1), we need only add the two individual transforms.

Theorem 3.2 (First Shifting Theorem). If the function $f(t)$ has a Fourier transform given by $F(w)$, then the Fourier transform of the function $f(t - a)$ is given by $F(w)e^{-2\pi iwa}$.

This theorem tells us that a phase shift in the time domain[†] gives rise to a sinusoidal type modulation in the frequency domain. The exact nature of this

---

[†]We call the space formed by $[w, F(w)]$ the frequency domain and the space $[t, f(t)]$ the temporal or time domain when the independent variable t represents time. When t represents a spatial variable, we call $[t, f(t)]$ the spatial domain.

modulation can be better appreciated if we write the Fourier transform in its rectangular complex form:

$$F(w) = F_R(w) + iF_I(w)$$

where $F_R(w)$ is the real portion and $F_I(w)$ is the imaginary portion of $F(w)$. In this form we have

$$F(w)e^{-2\pi iwa} = \left[F_R(w) + iF_I(w)\right]\left[\cos 2\pi wa - i \sin 2\pi wa\right],$$

or

(3.13)
$$\begin{aligned} F(w)e^{-2\pi iwa} = {} & F_R(w)\cos 2\pi wa + F_I(w)\sin 2\pi wa \\ & + i\left[F_I(w)\cos 2\pi wa - F_R(w)\sin 2\pi wa\right]. \end{aligned}$$

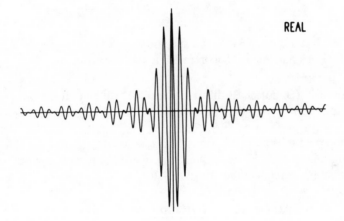

REAL

IMAGINARY

FIGURE 3.6. Transform of shifted pulse.

Thus we see that if $f(t)$ has a Fourier transform given by $F(w) = F_R(w) + iF_I(w)$, then the Fourier transform of $f(t - a)$ is formed from $F(w)$ as per Equation (3.13). We note that this modulation is such that the modulus of the transform is unchanged or invariant;

$$\|F(w)e^{-2\pi iwa}\| = \|F(w)\|\|e^{-2\pi iwa}\| = \|F(w)\|.$$

When $F(w)$ is real, then $F_I(w) = 0$ and Equation (3.13) yields

$$(3.14) \qquad F(w)e^{-2\pi iwa} = F_R(w)\cos 2\pi wa - iF_R(w)\sin 2\pi wa.$$

For example, the Fourier transform of the shifted pulse function $p_a(t - \alpha)$ is given by

$$G(w) = 2a\,\text{sinc}(2\pi wa)(\cos 2\pi w\alpha - i\sin 2\pi w\alpha).$$

This is shown in Figure 3.6 (using a specific value of α).

The dual or analogue of Theorem 3.2 is the following.

Theorem 3.3 (Second Shifting Theorem). If the function $f(t)$ has a Fourier transform given by $F(w)$, then the function $g(t) = f(t)e^{2\pi iat}$ has a Fourier transform given by $F(w - a)$.

This theorem tells us that a sinusoidal type modulation in the time domain results in a phase shift in the frequency domain. Let us now rewrite the results of Theorem 3.3 in rectangular form:

$$\mathcal{F}[f(t)\cos 2\pi at + if(t)\sin 2\pi at] = F(w - a).$$

When $-a$ is substituted for a in this equation we obtain

$$\mathcal{F}[f(t)\cos 2\pi at - if(t)\sin 2\pi at] = F(w + a).$$

Addition and subtraction of the preceding equations yields the results that are summarized as follows.

Theorem 3.4. If the function $f(t)$ has a Fourier transform given by $F(w)$ then

(i) $\quad \mathcal{F}[f(t)\cos 2\pi at] = \dfrac{F(w + a) + F(w - a)}{2}.$

(ii) $\quad \mathcal{F}[f(t)\sin 2\pi at] = i\dfrac{F(w + a) - F(w - a)}{2}.$

We now present a very useful theorem concerning scale change.

Theorem 3.5 (Scale Change). If the Fourier transform of $f(t)$ is $F(w)$, then the Fourier transform of $f(at)$ is given by

$$\frac{1}{|a|}F\left(\frac{w}{a}\right),$$

where a is any real number not equal to 0.

The obvious consequence of this theorem is the following.

Corollary 3.5. If the function $f(t)$ has a Fourier transform given by $F(w)$, then the Fourier transform of $f(-t)$ is given by $F(-w)$.

At this point even the most casual observer should have noticed the dual nature of or similarities between the Fourier transform pairs. The first and second shifting theorems illustrate this rather clearly. That is, a phase shift in the time domain gives rise to a sinusoidal type amplitude modulation in the frequency domain and a sinusoidal type amplitude modulation in the time domain results in a phase shift in the frequency domain. The next theorem amplifies these similarities and discusses the Fourier transform of a Fourier transform.

Theorem 3.6 (Transform of a Transform). If the function $f(t)$ has a Fourier transform given by $F(w)$, then

$$\mathscr{F}\{\mathscr{F}[f(t)]\} = f(-t).$$

This theorem tells us that if we perform the Fourier transform operation [Equation (3.1)] twice on a function $f(t)$, then we obtain the rotated function $f(-t)$. To illustrate the application of this theorem we will determine the Fourier transform of

$$\text{sinc } t = \frac{\sin t}{t}.$$

We first recall that the transform of a pulse function $p_a(w)$ is given by

$$\mathscr{F}[p_a(w)] = 2a\frac{\sin 2\pi at}{2\pi at}.$$

We now use Theorem 3.6 as follows:

$$\mathscr{F}\left[2a\frac{\sin 2\pi at}{2\pi at}\right] = \mathscr{F}\{\mathscr{F}[p_a(w)]\} = p_a(-w) = p_a(w).$$

Using the linearity Theorem 3.1 we have

$$\mathcal{F}\left[2a\frac{\sin 2\pi at}{2\pi at}\right] = 2a\mathcal{F}\left[\frac{\sin 2\pi at}{2\pi at}\right] = p_a(w).$$

thus

$$\mathcal{F}\left[\frac{\sin 2\pi at}{2\pi at}\right] = \frac{p_a(w)}{2a}.$$

Now, finally we use the change of scale Theorem 3.5 with scale factor $1/(2\pi a)$ to arrive at

$$\mathcal{F}\left[\frac{\sin t}{t}\right] = \pi p_a(2\pi aw) = \pi p_{1/2\pi}(w).$$

The next theorems discuss both the Fourier transform of the derivative of a function and the derivative of the Fourier transform of a function.

Theorem 3.7 (Derivative of the Transform). If both Fourier transforms $\mathcal{F}[f(x)] = F(w)$ and $\mathcal{F}[x^n f(x)]$ exist, then they are related as follows:

$$(-2\pi i)^n \mathcal{F}[x^n f(x)] = \frac{d^n F(w)}{dw^n}.$$

This theorem tells us that if the Fourier transform of $f(x)$ is $F(w)$, then the Fourier transform of $x^n f(x)$ is obtained by taking the nth derivative of $F(w)$ and dividing the result by $(-2\pi i)^n$. The dual or analogue of this theorem is the following.

Theorem 3.8 (Transform of the Derivative). If both Fourier transforms $\mathcal{F}[f(x)] = F(w)$ and $\mathcal{F}\{f^{[n]}(x)\}$ exist, then they are related as follows:

$$\mathcal{F}\{f^{[n]}(x)\} = (2\pi i w)^n F(w).$$

This theorem tells us that if the Fourier transform of $f(x)$ is $F(w)$, then the Fourier transform of the nth derivative of $f(x)$ is obtained by multiplying $F(w)$ by $(2\pi i w)^n$.

THE GAUSSIAN FUNCTION

The Gaussian function (shown in Figure 3.7) is described mathematically as

(3.15) $$f(x) = e^{-ax^2}.$$

This function is quite common in statistics and probability theory and is also known as the normal distribution function. An interesting property of this Gaussian function is that its Fourier transform is also a Gaussian function. This property is known as self-reciprocity. We now consider the Fourier transform of the Gaussian function.

By definition we have

$$F(w) = \int_{-\infty}^{\infty} e^{-ax^2} e^{-2\pi i w x} \, dx$$

$$= \int_{-\infty}^{\infty} e^{-a(x^2 + 2\pi i w x/a)} \, dx.$$

If we now write $(x^2 + 2\pi i w x/a)$ as $(x + i\pi w/a)^2 + (\pi w/a)^2$, then we have

$$F(w) = e^{-\pi^2 w^2/a} \int_{-\infty}^{\infty} e^{-a(x + i\pi w/a)^2} \, dx.$$

Next we make the change of variable $s = x + i\pi w/a$ which implies $ds = dx$

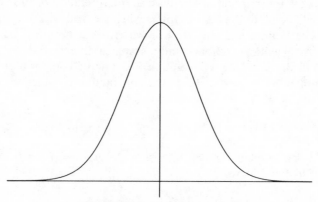

FIGURE 3.7. Gaussian function.

(since $dw = 0$), and we obtain

$$F(w) = e^{-\pi^2 w^2/a} \int_{-\infty}^{\infty} e^{-as^2}\, ds.$$

Inasmuch as s is a complex variable, this integral must be evaluated using contour integration. It can be shown (see, e.g., p. 239 of Raven's text listed in the Bibliography) that the value of this integral is $(\pi/a)^{1/2}$. Thus

(3.16)
$$\mathscr{F}\left[e^{-ax^2}\right] = \sqrt{\frac{\pi}{a}}\, e^{-\pi^2 w^2/a}.$$

When $a = \pi$, then we have complete self-reciprocity:

$$\mathscr{F}\left[e^{-\pi x^2}\right] = e^{-\pi w^2}.$$

CONVOLUTION AND CROSS-CORRELATION

The convolution of two functions $f(x)$ and $g(x)$, denoted as $f(x) * g(x)$, is defined as follows:

(3.17)
$$f(x) * g(x) = \int_{-\infty}^{\infty} f(\xi) g(x - \xi)\, d\xi.$$

The concept of convolution is inherent in almost every field of the physical sciences and engineering. For example, in mechanics it is known as the superposition or Duhamel integral. In systems theory it plays a crucial role as the impulse response integral, and in optics as the point spread or smearing function. A real appreciation of the physical interpretation of this convolution integral should come in the later chapters when we deal with specific physical applications of Fourier analysis. Here, however, we concentrate our effort on an applied mathematical approach and a graphical interpretation.

Mathematically speaking, convolution is a law of composition that combines two functions to yield a third. The convolution of functions is associative:

$$f(x) * [g(x) * h(x)] = [f(x) * g(x)] * h(x).$$

Convolution is also commutative; that is, $f(x) * g(x) = g(x) * f(x)$. This can be demonstrated as follows:

$$f(x) * g(x) = \int_{-\infty}^{\infty} f(\xi) g(x - \xi)\, d\xi.$$

If we now let $x - \xi = y$ (which implies $d\xi = -dy$), then we have

$$f(x) * g(x) = -\int_{\infty}^{-\infty} f(x - y) g(y) \, dy$$

$$f(x) * g(x) = \int_{-\infty}^{\infty} g(y) f(x - y) \, dy = g(x) * f(x).$$

Finally, convolution is distributive with respect to addition:

$$f(x) * [g(x) + h(x)] = f(x) * g(x) + f(x) * h(x),$$

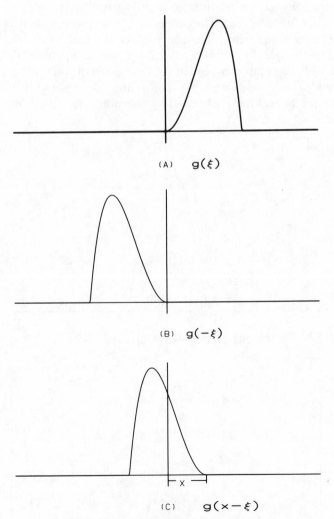

(A) $g(\xi)$

(B) $g(-\xi)$

(C) $g(x - \xi)$

FIGURE 3.8. Graphical representation of $g(x - \xi)$.

and

$$[g(x) + h(x)] * f(x) = g(x) * f(x) + h(x) * f(x).$$

We now present a graphical interpretation of the convolution integral. From Equation (3.17) we see that the convolution of $f(x)$ and $g(x)$ is given as the integral of the product of two functions $f(\xi)$ and $g(x - \xi)$. Let us first consider a graphical representation of $g(x - \xi)$. Shown in Figure 3.8a is the function $g(\xi)$. In Figure 3.8b we have rotated this function about the origin to obtain $g(-\xi)$, and in Figure 3.8c we have displaced this rotated function to the right (for positive values of x) by an amount x to obtain $g(x - \xi)$.

In Figure 3.9 we show this rotated and displaced function $g(x - \xi)$ along with the other function $f(\xi)$. The product of these two functions is shown as the dashed line in the figure. The convolution of f and g (for this particular value of x) is the area under this product curve [see Equation 3.17)].

Figure 3.10 shows this convolution for three different values of x. As can be appreciated from these figures, the convolution $f(x) * g(x) = h(x)$ for all values of x on the real line is obtained by sweeping this rotated function from $-\infty$ to ∞.

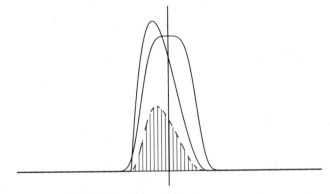

FIGURE 3.9.   Product of $f(\xi)$ and $g(x - \xi)$.

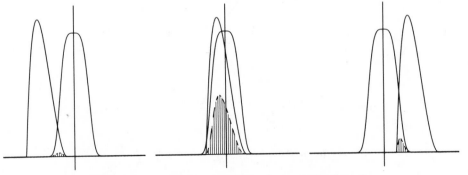

FIGURE 3.10.   Convolution integral for three specific values of x.

As an actual numerical example we now calculate the convolution of two equal pulse functions $[p_a(x)]$. Application of Equation (3.17) yields

$$h(x) = \int_{-\infty}^{\infty} p_a(\xi) p_a(x - \xi) \, d\xi$$

The task of evaluating this integral lies in determining the proper limits of integration. We have four possible situations shown in Figure 3.11a–3.11d and

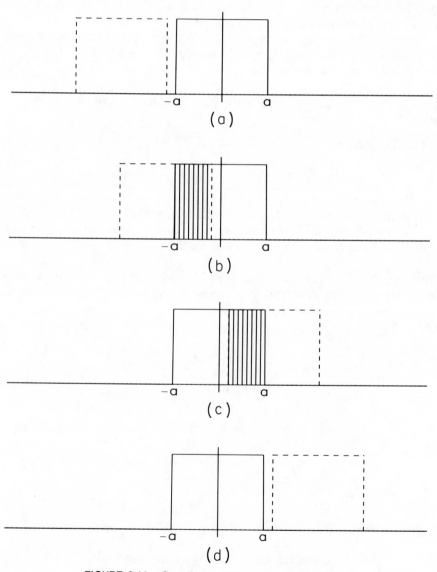

FIGURE 3.11.   Convolution of two equal pulse functions.

summarized as:

1. $x < -2a.$
2. $-2a \leqslant x < 0.$
3. $0 \leqslant x \leqslant 2a.$
4. $x > 2a.$

In the figures $f(\xi) = p_a(\xi)$ is shown as the solid line and the rotated and displaced function $g(x - \xi) = p_a(x - \xi)$ as the dashed line. The convolution (or area) is illustrated as the shaded region.

For cases 1 and 4 the two functions do not overlap and the product (and, consequently, the convolution) of these two functions is zero.

For case 2 the integral becomes

$$h(x) = \int_{-a}^{x+a} d\xi = x + 2a, \qquad -2a \leqslant x \leqslant 0.$$

Similarly for case 3 we have

$$h(x) = \int_{x-a}^{a} d\xi = 2a - x, \qquad 0 \leqslant x \leqslant 2a.$$

In Figure 3.12 we show this convolution function $h(x)$.

We now consider the Fourier transform of the convolution of two functions, which turns out to be a surprisingly nice result.

Theorem 3.9 (Convolution Theorem).   If both $f(x)$ and $g(x)$ have Fourier transforms given by $F(w)$ and $G(w)$, respectively, then the Fourier transform of $h(x) = f(x) * g(x)$ is given by $F(w)G(w)$.

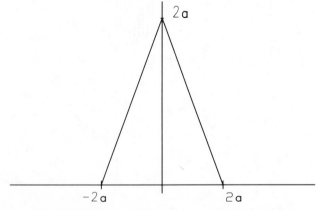

FIGURE 3.12.   Convolution of two equal pulse functions.

This theorem tells us that the Fourier transform of the convolution of two functions is simply the product of the two individual transforms. The analogue of this theorem is the product theorem which tells us that the Fourier transform of the product of two functions is the convolution of the two individual transforms. This is stated more formally in the following theorem.

Theorem 3.10 (Product Theorem). If both $f(x)$ and $g(x)$ have Fourier transforms given by $F(w)$ and $G(w)$, respectively, then the Fourier transform of the product of these two functions $h(x) = f(x)g(x)$ is given by $H(w) = F(w) * G(w)$.

We point out here that in the previous results the functions f and g (as well as F and G) may be complex, in which case the product $f(\xi)g(x - \xi)$ must be performed according to the product law for the complex number system.

We now consider a slightly different composition of two functions known as *cross-correlation*, which is defined as

$$(3.18) \qquad h(x) = f(x) \star g(x) = \int_{-\infty}^{\infty} f(\xi)g^*(\xi + x) \, d\xi.$$

When we formed the convolution (Equation 3.17) of two functions f and g we first rotated the second function g and then displaced it by an amount x. As can be seen from Equation (3.18), when we form the cross-correlation of two functions f and g we first take the complex conjugate of the second function g and then displace it by an amount $-x$. Thus for convolution we rotate the second function, whereas for cross-correlation we take the complex conjugate of the second function. The function $h(x) = f(x) \star g(x)$ is known as the cross-correlation function. It is important to note that this function is the result of applying the cross-correlation integral and, like the convolution integral, it is associative and distributive with respect to addition. Unlike convolution, however, *it is not commutative*; that is, $f(x) \star g(x) \neq g(x) \star f(x)$.

When a function $f(x)$ is cross-correlated with itself it is known as the *autocorrelation integral* and the resulting function, the *autocorrelation function*. Mathematically, the autocorrelation of f with itself is given as

$$(3.19) \qquad f(x) \star f(x) = \int_{-\infty}^{\infty} f(\xi)f^*(\xi + x) \, d\xi.$$

The following theorem tells us how to calculate the Fourier transform of a function that is the cross-correlation of two other functions.

Theorem 3.11. If both functions $f(x)$ and $g(x)$ have Fourier transforms given by $F(w)$ and $G(w)$, respectively, then the cross-correlation function $h(x) = f(x) \star g(x)$ has a Fourier transform given by $F(w)G^*(w)$ where $G^*(w)$ is the complex conjugate function of $G(w)$.

In the situation when f is cross-correlated with itself (i.e., the autocorrelation function) we have

(3.20) $$\mathcal{F}[f(x) \star f(x)] = F(w)F^*(w) = \|F(w)\|^2.$$

Now we write

$$h(x) = f(x) \star f(x) = \mathcal{F}^{-1}[F(w)F^*(w)],$$

or

$$\int_{-\infty}^{\infty} f(\xi)f^*(\xi - x)\,d\xi = \int_{-\infty}^{\infty} F(w)F^*(w)e^{2\pi i w x}\,dw.$$

If we let $x = 0$ in the preceding equation we obtain Parseval's energy formula:

$$\int_{-\infty}^{\infty} \|f(\xi)\|^2\,d\xi = \int_{-\infty}^{\infty} \|F(w)\|^2\,dw.$$

We now present a rather interesting example illustrating an application of the convolution theorem in which we determine the convolution integral of $e^{-\alpha x^2}$ and $e^{-\beta x^2}$:

$$h(x) = e^{-\alpha x^2} * e^{-\beta x^2} = \int_{-\infty}^{\infty} e^{-\alpha \xi^2}e^{-\beta(x-\xi)^2}\,d\xi.$$

Using Equation (3.16) we have

$$\mathcal{F}[e^{-\alpha x^2}] = \sqrt{\frac{\pi}{\alpha}}\, e^{-\pi^2 w^2/\alpha},$$

and

$$\mathcal{F}[e^{-\beta x^2}] = \sqrt{\frac{\pi}{\beta}}\, e^{-\pi^2 w^2/\beta}.$$

By the convolution theorem 3.9 we have

$$H(w) = \mathcal{F}[h(x)] = \sqrt{\frac{\pi^2}{\alpha\beta}}\, e^{-\pi^2 w^2(\alpha+\beta)/\alpha\beta}.$$

Finally, taking the inverse Fourier transform of this result we obtain

$$h(x) = \mathcal{F}^{-1}[H(w)] = \sqrt{\frac{\pi}{\alpha + \beta}}\, e^{-\alpha\beta x^2/(\alpha+\beta)}.$$

SYMMETRY RELATIONS

In this section we discuss various forms of symmetry of the Fourier transform. These results are often useful when dealing with (Fourier transform) frequency domain plots. The first theorem deals with the Fourier transform of the complex conjugate of a function.

Theorem 3.12. If the function $f(x)$ has a Fourier transform given by $F(w)$, then the Fourier transform of $f^*(x)$ [complex conjugate of $f(x)$] is given by $F^*(-w)$.

The next two theorems deal with symmetry when either the function or its transform is a pure real or a pure imaginary function.

Theorem 3.13. Assume that the function $f(x)$ has a Fourier transform given by $F(w)$, then:
  (i) If f is a real function then $F(w) = F^*(-w)$.
  (ii) If F is a real function then $f(x) = f^*(-x)$.

Theorem 3.14. Assume the function f has a Fourier transform given by $F(w)$, then
  (i)   If f is real and even, then $F(w)$ is real and even.
  (ii) If f is real and odd, then $F(w)$ is pure imaginary and odd.
  (iii) If f is pure imaginary and even, then $F(w)$ is pure imaginary and even.
  (iv) If f is pure imaginary and odd, then $F(w)$ is real and odd.

THE IMPULSE FUNCTION

Physically, an impulse is considered to be an extremely brief, very intense, unit area pulse. This concept is quite common, for example, in physics when we discuss such quantities as point charges, point masses, point sources, and concentrated forces. In this section we present a (nonrigorous) mathematical description of an impulse. There are several ways in which an impulse function can be defined. In each method the impulse is obtained as the limit of a sequence of well-behaved functions. Let us consider the sequence of pulse functions $\{p_{1/n}(x)\}$ where

$$p_{1/n}(x) = \begin{cases} n/2, & |x| \leqslant 1/n, \\ 0, & \text{otherwise.} \end{cases}$$

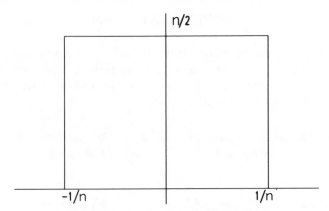

FIGURE 3.13. Generating pulse function.

A term of this sequence is shown in Figure 3.13. We note that the area of this pulse is $(n/2)(2/n) = 1$. Mathematically, we write

$$\int_{-\infty}^{\infty} p_{1/n}(x)\, dx = \int_{-1/n}^{1/n} \frac{n}{2}\, dx = 1 \quad \text{for all values of } n.$$

We now define the impulse, or delta function, as the limit of this sequence of unit area impulse functions:

$$\delta(x) = \lim_{n \to \infty} p_{1/n}(x).$$

This limiting process is illustrated in Figure 3.14 for $n = 1, 2, 3$. From the figure we can see that as n increases, the width of the pulse decreases toward zero, whereas the height or intensity increases without bound. However, the

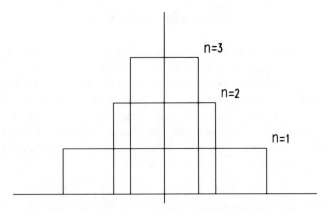

FIGURE 3.14. Limiting definition of the impulse function.

area (or strength) of each pulse remains equal to unity. Thus it seems reasonable to write

$$\int_{-\infty}^{\infty} \delta(x)\, dx = \int_{-\infty}^{\infty} \lim_{n \to \infty} p_{1/n}(x)\, dx = \lim_{n \to \infty} \int_{-\infty}^{\infty} p_{1/n}(x)\, dx = 1.$$

Earlier in this chapter we showed that the Fourier transform of $p_a(t)$ was given by $2a \operatorname{sinc} 2\pi wa$. From this we can infer that the Fourier transform of each term of the sequence $p_{1/n}(x)$ is given by $\mathscr{F}[p_{1/n}(x)] = \sin(2\pi w/n)/(2\pi w/n)$. From this we deduce

$$\mathscr{F}[\delta(x)] = \mathscr{F}\left[\lim_{n \to \infty} p_{1/n}(x) \right] = \lim_{n \to \infty} \mathscr{F}\left[p_{1/n}(x) \right]$$

$$= \lim_{n \to \infty} \frac{\sin 2\pi w/n}{2\pi w/n} = 1.$$

Thus we have "demonstrated" that the Fourier transform of the impulse function is the constant function $F(w) = 1$. The intent of this presentation is not to give a solid mathematical description of an impulse function and its Fourier transform. Instead, we are interested in providing the reader with a "physical feeling" of what an impulse is. Even though the approach that we take here would not be able to withstand any reasonable mathematical scrutiny, the end results are indeed valid and they can be rigorously demonstrated using the theory of distributions. We summarize our results in the following theorem.

Theorem 3.15. If $\delta(t)$ is a unit impulse function, then

$$\int_{-\infty}^{\infty} \delta(t)\, dt = \int_{0^-}^{0^+} \delta(t)\, dt = 1,$$

$$\mathscr{F}[\delta(t)] = 1.$$

In the remainder of this section we show how the results of this theorem can be used to obtain other interesting results. We first demonstrate that the Fourier transform of the unit constant function $[f(x) = 1]$ is a delta function. To do this we use Theorem 3.6 (transform of a transform) and write

(3.21) $$\mathscr{F}[\mathscr{F}[\delta(x)]] = \mathscr{F}[1] = \delta(-x) = \delta(x).$$

From Theorem 3.4 we recall that if $\mathscr{F}[f(t)] = F(w)$, then

$$\mathscr{F}[f(t)\cos 2\pi at] = \frac{F(w+a) + F(w-a)}{2},$$

$$\mathscr{F}[f(t)\sin 2\pi at] = i\frac{F(w+a) - F(w-a)}{2}.$$

If we let $f(t) = 1$ and use Equation (3.21), we find

(3.22) $$\mathscr{F}[\cos 2\pi at] = \frac{\delta(w+a) + \delta(w-a)}{2},$$

(3.23) $$\mathscr{F}[\sin 2\pi at] = i\frac{\delta(w+a) - \delta(w-a)}{2}.$$

Thus we see that the Fourier transform of a cosine function of frequency a is simply two delta functions of magnitude $\frac{1}{2}$ located at $w = -a$ and $w = a$. (Similar remarks hold for the Fourier transform of the sine function.)

We now consider the convolution of a function $f(x)$ with the delta function:

$$f(x) * \delta(x) = \int_{-\infty}^{\infty} \delta(\xi)f(x-\xi)\,d\xi = \int_{0^-}^{0^+} \delta(\xi)f(x-\xi)\,d\xi.$$

Inasmuch as $f(x-\xi)$ is basically a constant equal to $f(x)$ over the interval $0^- \leqslant \xi \leqslant 0^+$, we will remove it from under the integral sign to obtain

$$f(x) * \delta(x) = f(x)\int_{0^-}^{0^+} \delta(x)\,dx = f(x).$$

Thus we see that when we convolve a function $f(x)$ with the delta function we obtain the original function back. We can also show that

(3.24) $$f(x) * \delta(x-a) = f(x-a).$$

We have previously demonstrated that the Fourier transform of a pure cosine function $\cos 2\pi at$ is given by two delta functions of strength $\frac{1}{2}$ located at frequencies $-a$ and a. This is illustrated in Figure 3.15. As can be seen from this figure, the pure cosine function (which continues forever in the time domain) has single or concentrated frequencies of $w = +a$ and $w = -a$. Let us now consider the case when this cosine function is truncated as shown in Figure 3.16. Mathematically, we can express this function as the product of the pure cosine function $\cos 2\pi at$ and a pulse function of half-width b:

(3.25) $$f(x) = p_b(t)\cos 2\pi at.$$

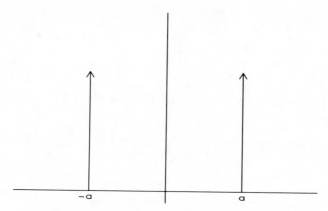

FIGURE 3.15. Fourier transform of the pure cosine function.

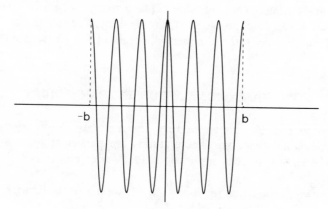

FIGURE 3.16. Truncated cosine function.

Using the product Theorem 3.10 we determine the Fourier transform of $f(x)$ to be

$$F(w) = \left[\frac{\delta(w + a) + \delta(w - a)}{2}\right] * 2b \sin c\, 2\pi wb.$$

Using Equation (3.24) we obtain

$$F(w) = b \sin c\,[2\pi b(w + a)] + b \sin c[2\pi b(w - a)].$$

This is shown in Figure 3.17.

As can be seen from the figure, when we truncate the pure (or single frequency) cosine function the effect is to add additional frequency components centered about $w = a$ and $w = -a$. The smaller the pulse width $2b$, the

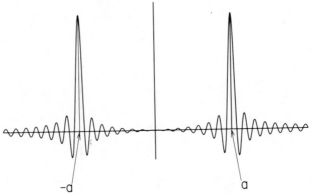

FIGURE 3.17. Fourier transform of the truncated cosine function.

more frequency components are added. These results should be compared to the illustrative example problem at the end of Chapter 2 in which we considered the (Fourier series) frequency content of the function $\sin 2\pi at$.

TWO-DIMENSIONAL FOURIER TRANSFORMS

In the remainder of this chapter we very briefly discuss the Fourier transform of two-dimensional functions. These results are particularly useful to us in Chapter 8 when we discuss applications of Fourier analysis to the field of optics.

Given a two-dimensional function $f(x, y)$, we define its Fourier transform pair[†] as

(3.26) $$F(u, v) = \int\int_{-\infty}^{\infty} f(x, y)e^{-2\pi i(ux+vy)} \, dx \, dy,$$

(3.27) $$f(x, y) = \int\int_{-\infty}^{\infty} F(u, v)e^{2\pi i(ux+vy)} \, du \, dv.$$

Just as in the one-dimensional case, Equation (3.26) is called the (direct) Fourier transform and Equation (3.27) is known as the inverse Fourier transform. We say $F(u, v)$ is the Fourier transform of $f(x, y)$ and that $f(x, y)$ is the inverse Fourier transform of $F(u, v)$. Notationally, we write

$$F(u, v) = \mathscr{F}[f(x, y)],$$
$$f(x, y) = \mathscr{F}^{-1}[F(u, v)].$$

[†] The integration in Equations (3.26) and (3.27) is to be carried out over the entire two-dimensional space and both integrals range from $-\infty$ to ∞. For notational simplification we just place one upper and one lower limit on the integrals when this does not lead to confusion.

We make the very important assumption that the integrals of both these equations exist and that each equation implies the other. That is, if we obtain $F(u, v)$ from Equation (3.26), then Equation (3.27) will uniquely return $f(x, y)$. Similarly, if we obtain $f(x, y)$ using Equation (3.27), then Equation (3.26) will return $F(u, v)$ uniquely.

PROPERTIES IN TWO DIMENSIONS

In this section we present the properties of the two-dimensional Fourier transform that are of use to us later in the book. For the most part these properties carry over directly from the theorems concerning the one-dimensional case. Consequently, they are presented without proof or comment.

Theorem 3.16 (Linearity). If both functions $f_1(x, y)$ and $f_2(x, y)$ have Fourier transforms given by $F_1(u, v)$ and $F_2(u, v)$, respectively, then the function $g(x, y) = af_1(x, y) + bf_2(x, y)$ has a Fourier transform given by $aF_1(u, v) + bF_2(u, v)$.

Theorem 3.17 (First Shifting Theorem). If the function $f(x, y)$ has a Fourier transform given by $F(u, v)$, then the Fourier transform of the function $f(x - a, y - b)$ is given by

$$F(u, v)e^{-2\pi i(au + bv)} = F(u, v)e^{-2\pi iau}e^{-2\pi ibv}$$

Theorem 3.18 (Second Shifting Theorem). If the function $f(x, y)$ has a Fourier transform given by $F(u, v)$, then the function

$$g(x, y) = f(x, y)e^{2\pi i(ax + by)} = f(x, y)e^{2\pi iax}e^{2\pi iby}$$

has a Fourier transform given by $F(u - a, v - b)$.

Theorem 3.19 (Scale Change). If the function $f(x, y)$ has a Fourier transform given by $F(u, v)$, then the Fourier transform of $f(ax, by)$ is given by

$$\frac{1}{|a||b|}F\left(\frac{u}{a}, \frac{v}{b}\right).$$

Theorem 3.20 (Transform of a Transform). If the function $f(x, y)$, has a Fourier transform given by $F(u, v)$, then

$$\mathscr{F}[\mathscr{F}[f(x, y)]] = f(-x, -y).$$

SEPARABLE AND RADIAL SYMMETRIC FUNCTIONS

There are two special classes of functions that afford significant simplification in the calculation of their two-dimensional Fourier transform because they can be reduced to a one-dimensional case.

The first class that we discuss is separable functions. We say a two-dimensional function $f(x, y)$ is separable if it can be written as the product of two one-dimensional functions $h(x)$ and $g(y)$:

$$(3.28) \qquad f(x, y) = h(x)g(y).$$

If a function f is separable into two functions h and g and both possess Fourier transforms given by H and G, respectively, then the two-dimensional Fourier transform of f is simply the product of the two individual transforms:

$$(3.29) \qquad F(u, v) = H(u)G(v).$$

This is easily verified. By definition we have

$$F(u, v) = \int\!\!\int_{-\infty}^{\infty} f(x, y)e^{-2\pi i(ux+vy)}\, dx\, dy,$$

$$F(u, v) = \int\!\!\int_{-\infty}^{\infty} h(x)g(y)e^{-2\pi iux}e^{-2\pi ivy}\, dx\, dy,$$

$$F(u, v) = \int_{-\infty}^{\infty} h(x)e^{-2\pi iux}\, dx \int_{-\infty}^{\infty} g(y)e^{-2\pi ivy}\, dy.$$

$$F(u, v) = H(u)G(v).$$

This equation points out the obvious fact that the Fourier transform of a separable function is also separable.

As an example, let us calculate the Fourier transform of the two-dimensional rectangle function, denoted as $\text{rect}_{ab}(x, y)$, and defined by

$$(3.30) \qquad \text{rect}_{ab}(x, y) = \begin{cases} 1, & -a \leqslant x \leqslant a, -b \leqslant y \leqslant b, \\ 0, & \text{otherwise.} \end{cases}$$

With a little thought we can see that this rectangle function can be written as the product of two pulse functions:

$$(3.31) \qquad \text{rect}_{ab}(x, y) = p_a(x)p_b(y).$$

The calculation of the transform of this function is trivial. That is to say, using

Equation (3.29) we have

$$\mathcal{F}\left[\text{rect}_{ab}(x, y)\right] = \mathcal{F}\left[p_a(x)\right]\mathcal{F}\left[p_b(y)\right]$$

or

$$\mathcal{F}\left[\text{rect}_{ab}(x, y)\right] = (2a \text{ sinc } 2\pi ua)(2b \text{ sinc } 2\pi vb)$$

(3.32) $$\mathcal{F}\left[\text{rect}_{ab}(x, y)\right] = 4ab \text{ sinc } 2\pi au \text{ sinc } 2\pi bv.$$

The other simplifying class of two-dimensional functions are known as *radial symmetric functions*. They are expressed in terms of polar coordinates. We begin our discussion by converting a two-dimensional function in Cartesian coordinates $\hat{f}(x, y)$ to a two-dimensional function in polar coordinates $f(r, \theta)$. We do so by using the following transformation equations:

(3.33)
$$x = r \cos \theta,$$
$$y = r \sin \theta,$$

where r is called the magnitude or radius and θ the angle. For example, to convert $\hat{f}(x, y) = xy$ to polar coordinates we proceed as follows:

$$\hat{f} = xy = r^2 \cos \theta \sin \theta = \tfrac{1}{2}r^2 \sin 2\theta = f(r, \theta).$$

A two-dimensional function (in polar coordinates) is called *radially symmetric* if it is independent of the angle θ or, in other words, if it can be expressed as a function of the radius only. For example, the function $\hat{f}(x, y) = x^2 + y^2$ is radially symmetric because

$$\hat{f}(x, y) = x^2 + y^2 = r^2(\cos^2 \theta + \sin^2 \theta) = r^2 = f(r).$$

To consider the Fourier transform of a radially symmetric function we first convert the transform variables u and v to polar coordinates by means of the transformation equations:

(3.34)
$$u = w \cos \varphi,$$
$$v = w \sin \varphi.$$

Using the coordinate transformation given by Equations (3.33) and (3.34) in the definition of the two-dimensional Fourier transform Equation (3.26), we obtain

$$F(w, \varphi) = \int_0^\infty \int_{-\pi}^\pi f(r, \theta) e^{-2\pi i wr(\cos \varphi \cos \theta + \sin \varphi \sin \theta)} r \, d\theta \, dr.$$

We now assume that f is a radially symmetric function and, therefore, can be written as a function of r only. Also let us use the formula for the cosine of the difference of two angles [i.e., $\cos(\alpha - \beta) = \cos \alpha \cos \beta + \sin \alpha \sin \beta$]. Then the preceding equation becomes

$$F(w, \varphi) = \int_0^\infty rf(r) \int_{-\pi}^\pi e^{-2\pi i w r \cos(\theta - \varphi)} \, d\theta \, dr.$$

We can now use the integral definition of the zero order Bessel function:

$$J_0(x) = \frac{1}{2\pi} \int_{-\pi}^\pi e^{-ix\cos(\theta - \varphi)} \, d\theta.$$

to rewrite this equation as

(3.35) $$F(w) = 2\pi \int_0^\infty rf(r) J_0(2\pi w r) \, dr.$$

The zero order Bessel function is a special case of the nth order Bessel function whose series expansion (for integer values of n) is given as

(3.36) $$J_n(x) = \sum_{k=0}^\infty \frac{(-1)^k}{k!(n + k)!} \left(\frac{x}{2}\right)^{n+2k}, \qquad n = 0, 1, 2, \ldots .$$

These Bessel functions are often used in many fields of science and engineering. Like the trigonometric functions, they are common and important enough to be tabulated in most advanced mathematics books. Note that in Equation (3.35) F is only a function of the radius w (we were able to remove the angle φ in the integral) and thus the transform F is also a radially symmetric function. This special case of the two-dimensional Fourier transform is known as the Fourier–Bessel transform or, more commonly, the Hankel transform. The inverse Hankel transform is defined as

(3.37) $$f(r) = 2\pi \int_0^\infty wF(w) J_0(2\pi w r) \, dw.$$

As an example let us calculate the Hankel transform of the circ function which is defined as

(3.38) $$\text{circ}(r) = \begin{cases} 1. & r \leqslant a, \\ 0, & \text{otherwise.} \end{cases}$$

In Cartesian coordinates this function is expressed as

$$\text{circ}(x, y) = \begin{cases} 1, & x^2 + y^2 \leqslant a^2, \\ 0, & \text{otherwise.} \end{cases}$$

Using Equation (3.35) we determine the Hankel transform of this function to be described by

$$F(w) = 2\pi \int_0^a rJ_0(2\pi wr)\, dr.$$

This is a rather common integral whose value is given as

$$F(w) = \frac{aJ_1(2\pi wa)}{w}.$$

where J_1 is the first order Bessel function. Next let us consider a radially symmetric Dirac delta function located at $r = a$, that is, $\delta(r - a)$.

NOTE: This is an infinitely thin annular ring of radius a. Using Equation (3.35) we determine the Hankel transform of this function to be

$$F(w) = 2\pi \int_0^\infty r\delta(r - a)J_0(2\pi wr)\, dr = 2\pi aJ_0(2\pi wa).$$

CONVOLUTION IN TWO DIMENSIONS

The two-dimensional convolution integral is defined the same, and enjoys the same properties, as the one-dimensional convolution integral. We define the convolution of two functions $f(x, y)$ and $g(x, y)$ as

$$(3.39) \quad f(x, y)*g(x, y) = \int\int_{-\infty}^{\infty} f(\xi, \eta)g(x - \xi, y - \eta)\, d\xi\, d\eta.$$

Just as with the one-dimensional case, we note that two-dimensional convolution is a law of composition that combines two functions to yield a third function. This law is commutative:

$$f(x, y)*g(x, y) = g(x, y)*f(x, y).$$

It is also associative:

$$f(x, y)*[g(x, y)*h(x, y)] = [f(x, y)*g(x, y)]*h(x, y),$$

and distributive with respect to addition:

$$f(x, y)*[g(x, y) + h(x, y)] = f(x, y)*g(x, y) + f(x, y)*h(x, y)$$

$$[g(x, y) + h(x, y)]*f(x, y) = g(x, y)*f(x, y) + h(x, y)*f(x, y)$$

We now present the two-dimensional product and convolution theorems that are duals of each other.

Theorem 3.21 (Convolution Theorem). If both $f(x, y)$ and $g(x, y)$ have Fourier transforms given by $F(u, v)$ and $G(u, v)$, respectively, then the Fourier transform of $h(x, y) = f(x, y) * g(x, y)$ is given by $F(u, v)G(u, v)$.

Theorem 3.22 (Product Theorem). If both $f(x, y)$ and $g(x, y)$ have Fourier transforms given by $F(u, v)$ and $G(u, v)$, respectively, then the Fourier transform of the product of these functions $h(x, y) = f(x, y)g(x, y)$ is given by $F(u, v) * G(u, v)$.

The two-dimensional cross-correlation integral is defined as

(3.40)

$$h(x, y) = f(x, y) \star g(x, y) = \int\int_{-\infty}^{\infty} f(\xi, \eta)g^*(\xi + x, \eta + y)\, d\xi\, d\eta.$$

Just as in the one-dimensional case, two-dimensional cross-correlation is a law of composition that is associative and distributive with respect to addition. However, *it is not commutative*. The crosscorrelation of a function with itself is known as the *autocorrelation integral*. Mathematically, we have

$$(3.41) \quad f(x, y) \star f(x, y) = \int\int_{-\infty}^{\infty} f(\xi, \eta)f^*(\xi + x, \eta + y)\, d\xi\, d\eta.$$

Theorem 3.23. If both functions $f(x, y)$ and $g(x, y)$ have Fourier transforms given by $F(u, v)$ and $G(u, v)$, respectively, then the cross-correlation function $h(x, y) = f(x, y) \star g(x, y)$ has a Fourier transform given by $F(u, v)G^*(u, v)$.

Theorem 3.24 (Parseval's Energy Theorem). If the function $f(x, y)$ has a Fourier transform given by $F(u, v)$, then

$$\int\int_{-\infty}^{\infty} \|f(x, y)\|^2\, dx\, dy = \int\int_{-\infty}^{\infty} \|F(u, v)\|^2\, du\, dv.$$

SUMMARY

In this chapter we presented the Fourier transform pair from an applied mathematics point of view. Emphasis was placed on the frequency content concept of the Fourier transform and the mathematical properties and manipulation of the transform. At the end of the chapter we generalized the one-dimensional results to two-dimensions.

BIBLIOGRAPHY

Sokolinkoff, I. S., and R. M. Redheffer, *Mathematics of Physics and Modern Engineering*. McGraw-Hill, New York, 1958.

Wylie, Jr., C. R., *Advanced Engineering Mathematics*. McGraw-Hill, New York, 1960.

Goodman, J. W., *Introduction to Fourier Optics*. McGraw-Hill, New York, 1968.

Papoulis, A., *Systems and Transforms with Applications in Optics*. McGraw-Hill, New York, 1968.

Parrent, G. B., and B., J. Thompson, *Physical Optics Notebook*. S.P.I.E., Redondo Beach, Calif., 1969.

Jahnke, E., Ed., *Tables of Higher Functions*, 6th ed. McGraw-Hill, New York, 1960.

Bracewell, R., *The Fourier Transform and its Applications*. McGraw-Hill, New York, 1965.

Arsac, J., *Fourier Transforms and the Theory of Distributions*. Prentice-Hall, Englewood Cliffs, N.J., 1966.

Titchmarsh, E. C., *Fourier Transforms*. Clarendon Press, Oxford, 1948.

Raven, F. H., *Mathematics of Engineering Systems*. McGraw-Hill, New York, 1966.

Worsnop, B. L., *An Introduction to Fourier Analysis*. Wiley, New York, 1961.

4

THE DISCRETE
FOURIER TRANSFORM

In the previous chapters we considered the Fourier series and Fourier transform of various functions. We may consider the Fourier series to be an operation that takes a function $f(x)$ and returns a sequence (of coefficients) C_k. By the same token, the Fourier transform may be considered an operation that maps a function $f(x)$ to another function $F(w)$. To determine either the Fourier series or Fourier transform of a function we must evaluate an integral. Thus it is only possible to consider functions that can be described analytically and even then these functions must be relatively uncomplicated. In the real world we rarely find such nice functions and, therefore, must turn to a digital computer for help. A digital computer, however, does not digest functions but instead, sequences of numbers that represent functions. In other words, we digitize an arbitrary function to obtain a sequence of numbers that can be handled by a computer. In this chapter we consider the discrete Fourier transform[†] which is an operation that maps a sequence $\{f(k)\}$ to another sequence $\{F(j)\}$.

[†] The discrete Fourier transform is also known as the finite Fourier transform.

Nth ORDER SEQUENCES

A finite sequence of N terms, or Nth order sequence, is defined as a function whose domain is the set of integers $\{0, 1, 2, \ldots, N - 1\}$ and whose range is the set of terms $\{f(0), f(1), f(2), \ldots, f(N - 1)\}$. Formally, we have an Nth order sequence in the set of ordered pairs

$$\{[0, f(0)], [1, f(1)], [2, f(2)], \ldots, [N - 1, f(N - 1)]\}.$$

In this chapter we use a shorter notation and follow the common practice of denoting the sequence as $\{f(k)\}$ and the kth term of the sequence as $f(k)$. In our work the terms are, in general, complex numbers. To clarify the previous remarks let us consider the sequence $\{f(k)\}$ whose terms are defined as:

$$f(k) = \frac{1}{k + 1}, \qquad k = 0, 1, \ldots, N - 1.$$

Thus

$$\{f(k)\} = \left\{ 1, \frac{1}{2}, \frac{1}{3}, \ldots, \frac{1}{N} \right\}.$$

We should not be misled into believing that we must have a formula or equation for $f(k)$ in terms of k to define a sequence. For example, the sequence $\{f(k)\} = \{1, -3, 4, 7.8, -3, \pi, 41\}$ is an eighth order sequence with no specific formula that maps k to $f(k)$ for $k \in [0, 7]$.

We now present some basic algebraic rules for sequences. The sum of two Nth order sequences $\{f(k)\}$ and $\{g(k)\}$, denoted as $\{f(k)\} + \{g(k)\}$, is defined as the sequence whose terms are given by $f(k) + g(k)$ for $k \in [0, N - 1]$. In other words, we add the individual term, for example,

$$\{0, 1, 2\} + \{4, 2, 1\} = \{4, 3, 3\}.$$

Subtraction is the inverse of addition. That is, to subtract two sequences we subtract the individual terms. Note that addition and subtraction are only defined for sequences that have the same order. The product of two sequences $\{f(k)\}$ and $\{g(k)\}$, denoted as $\{f(k)\}\{g(k)\}$, is defined as the sequence whose terms are given by $f(k)g(k)$ for $k \in [0, N - 1]$. In other words, we multiply the individual terms, for example,

$$\{0, 1, 2\}\{4, 2, 1\} = \{0, 2, 2\}.$$

Division is the inverse of multiplication and thus to divide the sequence $\{f(k)\}$

by $\{g(k)\}$ we divide the individual terms:

$$\frac{f(k)}{g(k)}, \quad g(k) \neq 0, \quad k \in [0, N-1].$$

NOTE: In general, the terms of the sequences will be complex and, therefore, the algebraic rules for multiplication (and division) of complex numbers must be used (see Chapter 1).

To multiply a sequence by a constant (perhaps complex), we simply multiply each term of the sequence by the constant. For example,

$$3\{1, 2, 3\} = \{3, 6, 9\}.$$

THE DISCRETE FOURIER TRANSFORM

Given any bounded[†] Nth order sequence $\{f(k)\}$, the *discrete Fourier transform pair* is defined as

$$F(j) = \frac{1}{N} \sum_{k=0}^{N-1} f(k) e^{-2\pi i k j/N} \qquad j \in [0, N-1],$$

$$f(k) = \sum_{j=0}^{N-1} F(j) e^{2\pi i k j/N} \qquad k \in [0, N-1].$$

If, for notational simplicity, we define the *weighting Kernel* W_N as

$$W_N = e^{2\pi i/N},$$

then the preceding equations become

(4.1) $\qquad F(j) = \frac{1}{N} \sum_{k=0}^{N-1} f(k) W_N^{-kj}, \qquad j = 0, 1, \ldots, N-1,$

(4.2) $\qquad f(k) = \sum_{j=0}^{N-1} F(j) W_N^{kj}, \qquad k = 0, 1, \ldots, N-1.$

Equation (4.1) is called the (*direct*) *discrete Fourier transform* and Equation (4.2) is known as the *inverse discrete Fourier transform*.

[†] We say a sequence is bounded if all of its terms are finite valued.

The discrete Fourier transform is an operation that maps an Nth order sequence $\{f(k)\}$ to another Nth order sequence $\{F(j)\}$. For example, the second order sequence $\{f(k)\} = \{1, 2\}$ has a discrete Fourier transform given by

$$F(0) = \tfrac{1}{2}\big[f(0)W_2^{-00} + f(1)W_2^{-10}\big],$$

$$F(1) = \tfrac{1}{2}\big[f(0)W_2^{-01} + f(1)W_2^{-11}\big],$$

or

$$F(0) = \tfrac{1}{2}(1 + 2) = \tfrac{3}{2},$$

$$F(1) = \tfrac{1}{2}(1 - 2) = -\tfrac{1}{2}.$$

Thus

$$\{F(j)\} = \{\tfrac{3}{2}, -\tfrac{1}{2}\}.$$

Using the inverse transform operation of Equation (4.2) on this sequence $\{F(j)\}$ we obtain:

$$f(0) = \big[F(0)W_2^{00} + F(1)W_2^{01}\big] = \tfrac{3}{2} + (-\tfrac{1}{2}) = 1,$$

$$f(1) = \big[F(0)W_2^{10} + F(1)W_2^{11}\big] = \tfrac{3}{2} + (-1)(\tfrac{1}{2}) = 2,$$

or

$$\{f(k)\} = \{1, 2\}.$$

We have just demonstrated for this particular sequence that the discrete Fourier transform possesses complete reciprocity. As it turns out, this result is true in general. That is, if we obtain the sequence $\{F(j)\}$ from the sequence $\{f(k)\}$ using Equation (4.1), then Equation (4.2) will uniquely return $\{f(k)\}$ from $\{F(j)\}$. For the sake of convenient reference we will catalogue this result as a theorem.

Theorem 4.1 (Reciprocity). The discrete Fourier transform possesses complete reciprocity; that is, it is unique.

Notationally, we write

$$\{F(j)\} = \overline{\mathscr{F}}\big[\{f(k)\}\big]$$

$$\{f(k)\} = \overline{\mathscr{F}}^{-1}\big[\{F(j)\}\big],$$

to indicate that $\{f(k)\}$ and $\{F(j)\}$ are discrete Fourier transform pairs.

So far we have only considered the sequences $\{f(k)\}$ and $\{F(j)\}$ over the Nth order domain $[0, N - 1]$. We now wish to look at the behavior of $\{f(k)\}$ and $\{F(j)\}$ for values of k and j outside this domain. However, we must first consider some basic properties of the weighting kernel W_N.

$$(4.3) \qquad W_N^{nN} = e^{2\pi i n N/N} = e^{2\pi i n} = 1 \quad \text{for any integer } n,$$

$$(4.4) \qquad W_N^{(k+j)} = e^{2\pi i(k+j)/N} = e^{2\pi i k/N}e^{2\pi i j/N} = W_N^k W_N^j.$$

Using these two properties it is a trivial task to show that W_N is periodic in both k and j with period N:

$$(4.5) \qquad W_N^{(k+N)j} = W_N^{kj}W_N^{Nj} = W_N^{kj},$$

$$(4.6) \qquad W_N^{k(j+N)} = W_N^{kj}W_N^{kN} = W_N^{kj},$$

$$(4.7) \qquad W_N^{k(N-j)} = W_N^{kN}W_N^{-kj} = W_N^{-kj},$$

$$(4.8) \qquad W_N^{(N-k)j} = W_N^{Nj}W_N^{-kj} = W_N^{-kj}.$$

We are now ready to consider our first extension theorem.

Theorem 4.2 (Periodicity). The discrete Fourier transform $\{F(j)\}$ and the discrete inverse Fourier transform $\{f(k)\}$ are both periodic with periodicity N; that is, (i) $F(j + N) = F(j)$ and (ii) $f(k + N) = f(k)$.

Proof.

$$\text{(i)} \quad F(j + N) = \frac{1}{N}\sum_{k=0}^{N-1} f(k)W_N^{-(j+N)k} = \frac{1}{N}\sum_{k=0}^{N-1} f(k)W_N^{-Nk}W_N^{-kj},$$

$$F(j + N) = \frac{1}{N}\sum_{k=0}^{N-1} f(k)W_N^{-kj} = F(j).$$

$$\text{(ii)} \quad f(k + N) = \sum_{j=0}^{N-1} F(j)W_N^{(k+N)j} = \sum_{j=0}^{N-1} F(j)W_N^{kj}W_N^{Nj},$$

$$f(k + N) = \sum_{j=0}^{N-1} F(j)W_N^{kj} = f(k).$$

This theorem tells us that the discrete Fourier transform is periodic with periodicity N or, in other words, this theorem describes the behavior of $\{F(j)\}$

for $j \geqslant N$. We emphasize again that the transform as given by Equation (4.1) is only defined for values of j ranging from 0 to $N - 1$. However, if we use Equation (4.1) to calculate $F(j)$ for values of $j \geqslant N$, then we will still obtain a value and Theorem 4.2 tells us what that value will be. For example, we have already illustrated that the discrete Fourier transform of $\{f(k)\} = \{1, 2\}$ is given by $\{F(j)\} = \{\frac{3}{2}, -\frac{1}{2}\}$. If we now use Equation (4.1) to obtain $F(2)$ and $F(3)$ we find:

$$F(2) = \frac{1}{2}\left[f(0)W_2^{-02} + f(1)W_2^{-12}\right] = \frac{1}{2}(1 + 2) = \frac{3}{2},$$

$$F(3) = \frac{1}{2}\left[f(0)W_2^{-03} + f(1)W_2^{-13}\right] = \frac{1}{2}(1 - 2) = -\frac{1}{2}.$$

Now using Theorem 4.2 we find

$$F(2) = F(2 + 0) = F(0) = \frac{3}{2},$$

$$F(3) = F(2 + 1) = F(1) = -\frac{1}{2}.$$

Note that when we transform $\{f(k)\}$ as per Equation (4.1), we only use values of $\{f(k)\}$ for k between 0 and $N - 1$ and that $\{F(j)\}$ is absolutely independent of the behavior of $\{f(k)\}$ outside $[0, N - 1]$. However, if we ever have occasion to extend $\{f(k)\}$, we will use Theorem 4.2 as a guide and define $f(N + k) = f(k)$.

The next theorem tells us how to extend the sequences $\{F(j)\}$ and $\{f(k)\}$ outside the domain $[0, N - 1]$ for negative values of the indices j and k.

Theorem 4.3 (Negative Indices). If the sequences $\{f(k)\}$ and $\{F(j)\}$ are discrete Fourier transform pairs then

(i) $F(-j) = F(N - j)$, $j \in [0, N - 1]$,
(ii) $f(-k) = f(N - k)$, $k \in [0, N - 1]$.

Proof.
(a) Using Equations (4.1) and (4.7) we find

$$F(N - j) = \frac{1}{N}\sum_{k=0}^{N-1} f(k)W_N^{-k(N-j)}$$

$$= \frac{1}{N}\sum_{k=0}^{N-1} f(k)W_N^{-k(-j)} = F(-j).$$

(b) This proof is similar to that of part (a) and, therefore, is omitted.

We now have a meaning for negative indices, or a justification for considering $F(-j)$ and $f(-k)$. In our previous example we showed

$$\overline{\mathcal{F}}[\{1,2\}] = \{\tfrac{3}{2}, -\tfrac{1}{2}\}.$$

Now using Theorem 4.3 we are able to write

$$F(-1) = F(2-1) = F(1) = -\tfrac{1}{2}.$$

We would have obtained this same result if we used Equation (4.1) with $j = -1$.

PROPERTIES OF THE DISCRETE FOURIER TRANSFORM

Given any bounded sequence $\{f(k)\}$ we can always calculate its discrete Fourier transform using Equation (4.1). This calculation requires N^2 mathematical multiplications and additions, and for large values of N this can become unwieldy. In this section we present several theorems concerning the properties of the discrete Fourier transform that can often be used to simplify its calculation. Just as we did in the previous two chapters for the Fourier series and Fourier transform, we present these results as theorems because this approach affords the most compact and easily referenced way to catalogue our results.

Theorem 4.4 (Linearity).   If $\{F(j)\} = \overline{\mathcal{F}}[\{f(k)\}]$, $\{G(j)\} = \overline{\mathcal{F}}[\{g(k)\}]$, and $\{h(k)\} = a\{f(k)\} + b\{g(k)\}$, then $\{H(j)\} = \overline{\mathcal{F}}[\{h(k)\}] = a\{F(j)\} + b\{G(j)\}$.

This theorem tells us how to calculate the discrete Fourier transform of a sequence that is a linear combination of two other sequences whose Fourier transforms are known. For example, for the sixth order sequences

$$\{\delta(k)\} = \{1,0,0,0,0,0\},$$

$$\{f(k)\} = \{1,2,3,3,2,1\},$$

we can calculate

$$\overline{\mathcal{F}}[\{\delta(k)\}] = \tfrac{1}{6}\{1,1,1,1,1,1\},$$

$$\overline{\mathcal{F}}[\{f(k)\}] = \{(2,0),(-0.5,-0.288),(0,0),(0,0),(0,0),(-0.5,0.288)\}.$$

Thus if $\{g(k)\} = 6\{\delta(k)\} + \{f(k)\} = \{7, 2, 3, 3, 2, 1\}$, then

$$\{G(j)\} = \bar{\bar{\mathscr{F}}}[\{g(k)\}]$$

$$= \{(3, 0), (0.5, -0.288), (1, 0), (1, 0), (1, 0), (0.5, 0.288)\}.$$

NOTE: Since addition is defined only when the sequences have the same order, this theorem only applies to sequences that have the same order.

Theorem 4.5 (First Shifting Theorem).   If the discrete Fourier transform of the Nth order sequence $\{f(k)\}$ is $\{F(j)\}$, then the discrete Fourier transform of the shifted sequence $\{f(k - n)\}$, $n \in [0, N - 1]$, is given by $\{F(j)W_N^{-jn}\}$.

This theorem tells us that if we know the discrete Fourier transform of the sequence $\{f(k)\}$, then we can obtain the discrete Fourier transform of the shifted sequence $\{f(k - n)\}$ with only N additional multiplications. For example, let's begin with the sequence $\{f(k)\} = \{1, 2, 3, 3, 2, 1\}$. To obtain the shifted sequence $\{f(k - 3)\}$, we simply shift each term to the left by three and recall that Theorem 4.3 says that $\{f(k)\}$ may be considered periodic. Thus the terms that "fall off" on the left reappear on the right;

$$\{f(k - 3)\} = \{3, 2, 1, 1, 2, 3\}.$$

If we now use Theorem 4.5 to calculate the discrete Fourier transform of this sequence, we obtain

$$\bar{\bar{\mathscr{F}}}[\{f(k - 3)\}] = \{F(j)W_6^{-3j}\} = \{F(j)e^{-2\pi i 3j/6}\} = \{F(j)(-1)^j\},$$

$$\bar{\bar{\mathscr{F}}}[\{f(k - 3)\} = \{(2, 0), (0.5, 0.288), (0, 0), (0, 0), (0, 0), (0.5, -0.288)\}.$$

Theorem 4.6 (Second Shifting Theorem).   If the discrete Fourier transform of the Nth order sequence $\{f(k)\}$ is $\{F(j)\}$, then the discrete Fourier transform of the sequence $\{f(k)W_N^{nk}\}$ is given by $\{F(j - n)\}$, $n \in [0, N - 1]$.

Note the dual nature of the discrete Fourier transform illustrated by these two theorems. Theorem 4.5 tells us that a "phase" shift in the k (or time) domain gives rise to a sinusoidal type modulation in the frequency domain. Theorem 4.6 tells us that a sinusoidal type modulation in the k domain give rise to a "phase" shift in the frequency domain. Theorem 4.6 can also be used to obtain the following useful result.

Theorem 4.7. If the Nth order sequence $\{f(k)\}$ has the discrete Fourier transform $\{F(j)\}$ then

(i) $\qquad \overline{\mathscr{F}}\left[\left\{ f(k)\cos \dfrac{2\pi kn}{N} \right\}\right] = \dfrac{1}{2}[\{F(j+n) + F(j-n)\}],$

(ii) $\qquad \overline{\mathscr{F}}\left[\left\{ f(k)\sin \dfrac{2\pi kn}{N} \right\}\right] = \dfrac{i}{2}[\{F(j+n) - F(j-n)\}],$

$$n \in [0, N-1].$$

Proof. As a direct consequence of the previous theorem we have

$$\overline{\mathscr{F}}\left[\left\{ f(k)\cos \frac{2\pi kn}{N} + i f(k)\sin \frac{2\pi kn}{N} \right\}\right] = \{F(j-n)\},$$

$$\overline{\mathscr{F}}\left[\left\{ f(k)\cos \frac{2\pi kn}{N} - i f(k)\sin \frac{2\pi kn}{N} \right\}\right] = \{F(j+n)\}.$$

Addition of these two equations yields the results of part (i) whereas subtraction yields part (ii).

We know from a previous example that the discrete Fourier transform of the sequence $\{f(k)\} = \{1, 2, 3, 3, 2, 1\}$ is given by the sequence $\{F(j)\} = \{(2,0), (-0.5, -0.288), (0,0), (0,0), (0,0), (-0.5, -0.288)\}$. For $n = 3$ we have

$$\{f(k)\cos \pi k\} = \{1, -2, 3, -3, 2, -1\}$$

and using Theorem 4.7i we find

$$\overline{\mathscr{F}}[\{f(k)\cos \pi k\}] = \{(0,0), (0,0), (-0.5, 0.288), (2,0), (-0.5, -0.288), (0,0)\}.$$

We now present a theorem concerning the discrete Fourier transform of the discrete Fourier transform of a sequence.

Theorem 4.8 (Transform of a Transform). If the Nth order sequence $\{F(j)\}$ is the discrete Fourier transform of the sequence $\{f(k)\}$, then

$$\overline{\mathscr{F}}\left[\overline{\mathscr{F}}[\{f(k)\}]\right] = \frac{1}{N}\{f(-k)\} = \frac{1}{N}\{f(N-k)\}.$$

In Chapter 1 we defined the derivative of a function as

$$f'(x) = \lim_{h \to 0} \frac{f(x+h) - f(x)}{h}.$$

The analogous operation for sequences is known as the (forward) difference of a sequence and is defined as

$$\{\Delta f(k)\} = \{f(k + 1) - f(k)\}.$$

In Chapter 3 we presented theorems that related the derivative and the Fourier transform of a function. Next we present the analogous theorems concerning the difference of the discrete Fourier transform of a sequence and the discrete Fourier transform of the difference of a sequence.

Theorem 4.9 (Difference of the Transform). If $\{F(j)\}$ is the discrete Fourier transform of the Nth order sequence $\{F(k)\}$, then

$$\{\Delta F(j)\} = \overline{\mathscr{F}}\left[\{f(k)(W_N^{-k} - 1)\}\right].$$

Theorem 4.10 (Transform of the Difference). If $\{F(j)\}$ is the discrete Fourier transform of the Nth order sequence $\{f(k)\}$, then

$$\overline{\mathscr{F}}\left[\{\Delta f(k)\}\right] = \{F(j)(W_N^j - 1)\} = \{F(j)\}\{W_N^j - 1\}.$$

As an illustrative example let us determine the discrete Fourier transform of the sixth order sequence $\{g(k)\} = \{1, 1, 0, -1, -1, 0\}$. Closer examination of this sequence reveals that it is simply the forward difference of the sequence $\{f(k)\} = \{1, 2, 3, 3, 2, 1\}$:

$$\{g(k)\} = \{\Delta f(k)\}.$$

NOTE: Theorem 4.2 was used to determine $g(5)$; that is, $g(5) = f(6) - f(1)$ $= f(0) - f(1)$.

Now we can use Theorem 4.10 to write

$$\{G(j)\} = \{F(j)\}\{W_N^j - 1\},$$

$$\{G(j)\} = \{(3, 0), (-0.5, -0.288), (0, 0), (0, 0), (0, 0), (-0.5, 0.288)\}$$

$$\{(0, 0), (-0.5, 0.866), (-1.5, 0.866), (-2, 0),$$

$$(-1.5, -0.866), (-0.5, -0.866)\}$$

$$\{G(j)\} = \{(0, 0), (0.5, 0.288), (0, 0), (0, 0), (0, 0), (0.5, 0.288)\}.$$

Referring back to Chapter 3 in which we presented a discussion of Fourier transforms, we note a definite analogy between the theorems of that chapter and the theorems of this chapter. We point out here, however, that we do not have a theorem concerning the scale change of a sequence. If we did, it would have to read something like "if the discrete Fourier transform of the sequence $\{f(k)\}$ is given by $\{F(j)\}$, then for any integer $m \neq 0$, the discrete Fourier transform of the sequence $\{f(mk)\}$ is given by $\{F(j/m)/|m|\}$." Although this may at first appear valid, we must recall that an Nth order sequence is a function whose domain is a finite set of integers. However, in general j/m will not be an integer and thus $F(j/m)$ is meaningless. The closest we can come to presenting a scale change theorem is the following.

Theorem 4.11. If $\{F(j)\}$ is the discrete Fourier transform of the Nth order sequence $\{f(k)\}$, then the discrete Fourier transform of $\{f(-k)\}$ is given by $\{F(-j)\}$.

SYMMETRY RELATIONS

In this section we discuss various forms of symmetry for the discrete Fourier transform. These results can be particularly useful for calculational purposes.

We say that an Nth order sequence $\{f(k)\}$ is even if and only if

$$f(N - k) = f(-k) = f(k). \qquad k \in [0, N - 1].$$

Similarly, an Nth order sequence $\{f(k)\}$ is said to be odd if and only if

$$f(N - k) = f(-k) = -f(k), \qquad k \in [0, N - 1].$$

For example, the sequence $\{6, 1, 3, 3, 1\}$ is considered even and the sequence $\{1, 3, -3, 1\}$ is odd. The next theorem deals with odd and even sequences.

Theorem 4.12. Assume that $\{F(j)\}$ is the discrete Fourier transform of the Nth order sequence $\{f(k)\}$; then:

(i) $\{f(k)\}$ is an even sequence implies $\{F(j)\}$ is an even sequence.
(ii) $\{f(k)\}$ is an odd sequence implies $\{F(j)\}$ is an odd sequence.

A sequence is said to be real if and only if all of its terms are real valued. A sequence is said to be imaginary (or pure imaginary) if and only if all of its terms are imaginary valued. Given a sequence $\{f(k)\}$, its *complex conjugate sequence* [denoted as $\{f^*(k)\}$] is defined as the sequence whose terms are the complex conjugates of those of $\{f(k)\}$. The next theorems deal with symmetry of complex sequences and their transforms.

Theorem 4.13. Assume that $\{F(j)\}$ is the discrete Fourier transform of the Nth order sequence $\{f(k)\}$; then:

   (i) if $\{f(k)\}$ is a real sequence, then $\{F(j)\} = \{F^*(-j)\}$,
   (ii) if $\{F(j)\}$ is a real sequence, then $\{f(k)\} = \{f^*(-k)\}$.

Theorem 4.14. Assume that $\{F(j)\}$ is the discrete Fourier transform of the Nth order sequence $\{f(k)\}$; then:

   (i) if $\{f(k)\}$ is an imaginary sequence, then $\{F(j)\} = -\{F^*(-j)\}$,
   (ii) if $\{F(j)\}$ is an imaginary sequence, then $\{f(k)\} = -\{f^*(-k)\}$.

Theorem 4.15. Assume that $\{F(j)\}$ is the discrete Fourier transform of the Nth order sequence $\{f(k)\}$; then:

   (i)   if $\{f(k)\}$ is a real and even sequence, then $\{F(j)\}$ is a real and even sequence,
   (ii) if $\{f(k)\}$ is a real and odd sequence, then $\{F(j)\}$ is an imaginary and odd sequence,
   (iii) if $\{f(k)\}$ is an imaginary and even sequence, then $\{F(j)\}$ is an imaginary and even sequence.
   (iv) if $\{f(k)\}$ is an imaginary and odd sequence, then $\{F(j)\}$ is a real and odd sequence.

SIMULTANEOUS CALCULATION OF REAL TRANSFORMS

In this section we demonstrate how Theorems 4.4 and 4.13 may be combined to obtain an extremely useful computational technique. This technique permits the simultaneous calculation of the discrete Fourier transform of two real sequences. We begin by considering two real sequences $\{f_1(k)\}$ and $\{f_2(k)\}$ with discrete Fourier transforms $\{F_1(j)\}$ and $\{F_2(j)\}$, respectively. We first combine these two real sequences to obtain the complex sequence $\{f(k)\}$:

(4.9) $$\{f(k)\} = \{f_1(k)\} + i\{f_2(k)\}.$$

Using Theorem 4.4 (linearity) we have

(4.10) $$\{F(j)\} = \overline{\overline{\mathscr{F}}}\big[\{f(k)\}\big] = \{F_1(j)\} + i\{F_2(j)\}.$$

Taking the complex conjugate of both sides of this equation we obtain

(4.11) $$\{F^*(j)\} = \{F_1^*(j)\} - i\{F_2^*(j)\}.$$

We now substitute $N - j$ for j in Equation (4.11) and obtain

(4.12) $\{F^*(N - j)\} = \{F_1^*(N - j)\} - i\{F_2^*(N - j)\}.$

However, since $\{f_1(k)\}$ and $\{f_2(k)\}$ are real sequences, we know from Theorem 4.13 that

$$\{F_1(j)\} = \{F_1^*(-j)\} = \{F_1^*(N - j)\},$$

$$\{F_2(j)\} = \{F_2^*(-j)\} = \{F_2^*(N - j)\}.$$

Substituting these expressions into Equation (4.12) yields

(4.13) $\{F^*(N - j)\} = \{F_1(j)\} - i\{F_2(j)\}.$

Addition and subtraction of Equations (4.10) and (4.13) produces our desired result:

(4.14)

$$\{F_1(j)\} = \frac{\{F(j)\} + \{F^*(N - j)\}}{2},$$

$$\{F_2(j)\} = \frac{\{F(j)\} - \{F^*(N - j)\}}{2i}.$$

Thus we have just shown how to calculate the Fourier transform of two real sequences by using Equation (4.1) only once. For example, let us use this technique to calculate the discrete Fourier transform of the two fifth order sequences $\{f_1(k)\} = \{0, 1, 2, 9, 16\}$ and $\{f_2(k)\} = \{1, 2, 4, 8, 16\}$. We first combine these two sequences as per Equation (4.9) to obtain the complex sequence

$$\{f(k)\} = \{(0, 1), (1, 2), (2, 4), (9, 8), (16, 16)\},$$

and then use Equation (4.1) to determine that the discrete Fourier transform of $\{f(k)\}$ is given as

$$\{F(j)\} = \{(5.6, 6.2), (-3.862, 3.047), (-2.956, -1.539),$$

$$(-1.186, -2.403), (2.404, -4.305)\}.$$

We now use Equations (4.14) with $N = 5$ to extract the two individual transforms $\{F_1(j)\}$ and $\{F_2(j)\}$:

$$F_1(0) = \frac{F(0) + F^*(5)}{2} = \frac{(5.6, 6.2) + (5.6, -6.2)}{2} = (5.6, 0),$$

$$F_2(0) = \frac{F(0) - F^*(5)}{2i} = \frac{(5.6, 6.2) - (5.6, -6.2)}{2i} = \frac{(0, 6.2)}{i} = (6.2, 0),$$

$$F_1(1) = \frac{F(1) + F^*(4)}{2} = \frac{(-3.862, 3.047) + (2.404, 4.305)}{2} = (-0.729, 3.676),$$

$$F_2(1) = \frac{F(1) - F^*(4)}{2i} = \frac{(-3.862, 3.047) - (2.404, 4.305)}{2i} = (-0.629, 3.133).$$

Continuing in this way we obtain

$$\{F_1(j)\} = \{(5.6, 0), (-0.719, 3.676), (-1.070, 0.431),$$

$$(-1.070, -0.431), (-0.719, -3.676)\},$$

$$\{F_2(j)\} = \{(6.2, 0), (-0.629, 3.133), (-1.970, 0.885),$$

$$(-1.970, -0.885), (-0.629, -3.133)\}.$$

THE FAST FOURIER TRANSFORM

In 1965 J. W. Tukey and J. W. Cooley published an algorithm that under certain conditions, tremendously reduces the number of computations required to compute the discrete Fourier transform of a sequence. This algorithm has come to be called the fast Fourier transform (FFT) and is considered one of the most significant contributions to numerical analysis of this century. Basically, the fast Fourier transform is a clever computational technique of sequentially combining progressively larger weighted sums of data samples so as to produce the discrete Fourier transform. These comments become more apparent as we proceed.

Let us begin by assuming that we have an Nth order sequence $\{f(k)\}$ with discrete Fourier transform $\{F(j)\}$. Furthermore, we assume that N is an even integer and thus we can form the two new subsequences:

$$
\begin{aligned}
&f_1(k) = f(2k),\\
&(4.15)\\
&f_2(k) = f(2k + 1), \qquad k = 0,\ldots, M - 1, \text{ where } M = N/2.
\end{aligned}
$$

For example, if $\{f(k)\} = \{0, 1, 2, 14, 16, 20\}$, then $\{f_1(k)\} = \{0, 2, 16\}$ and $\{f_2(k)\} = \{1, 14, 20\}$. Note that

$$f_1(k + M) = f(2(k + M)) = f(2k + N) = f(2k) = f_1(k),$$

$$f_2(k + M) = f(2(k + M) + 1) = f(2k + N + 1) = f(2k + 1) = f_2(k),$$

and, therefore, we see that both $\{f_1(k)\}$ and $\{f_2(k)\}$ are periodic sequences with periodicity M.

Since $\{f_1(k)\}$ and $\{f_2(k)\}$ are Mth order sequences we can use Equation (4.1) to determine their discrete Fourier transforms:

(4.16)
$$F_1(j) = \frac{1}{M} \sum_{k=0}^{M-1} f_1(k) W_M^{-kj},$$
$$j = 0, \ldots, M - 1,$$
$$F_2(j) = \frac{1}{M} \sum_{k=0}^{M-1} f_2(k) W_M^{-kj}.$$

We note that by Theorem 4.2, both $\{F_1(j)\}$ and $\{F_2(j)\}$ are periodic with periodicity M. Now let us consider the discrete Fourier transform of the Nth order sequence $\{f(k)\}$:

(4.17)
$$F(j) = \frac{1}{N} \sum_{k=0}^{N-1} f(k) W_N^{-kj}.$$

Splitting the summation we can rewrite the preceding equation as

$$F(j) = \frac{1}{N} \sum_{k=0}^{M-1} f(2k) W_N^{-2kj} + \frac{1}{N} \sum_{k=0}^{M-1} f(2k + 1) W_N^{-(2k+1)j}.$$

However, we note that

$$W_N^{-2kj} = e^{-2\pi i k j/(N/2)} = W_M^{-kj},$$

$$W_N^{-(2k+1)j} = e^{-2\pi i j(2k+1)/N} = W_M^{-kj} W_N^{-j}.$$

Therefore, the previous equation becomes

$$F(j) = \frac{1}{N} \sum_{k=0}^{M-1} f_1(k) W_M^{-kj} + \frac{W_N^{-j}}{N} \sum_{k=0}^{M-1} f_2(k) W_M^{-kj},$$

$$j = 0, \ldots, N - 1.$$

Comparison to Equations (4.16) yields

(4.18) $$F(j) = \frac{F_1(j)}{2} + \frac{W_N^{-j}F_2(j)}{2}, \qquad j = 0,\ldots, N-1.$$

Because $\{F_1(j)\}$ and $\{F_2(j)\}$ are periodic (with period M) we have

(4.19)
$$F(j) = \tfrac{1}{2}\big[F_1(j) + F_2(j)W_N^{-j}\big],$$
$$j = 0,\ldots, M-1,$$
$$F(j+M) = \tfrac{1}{2}\big[F_1(j) - F_2(j)W_N^{-j}\big).$$

As we have noted, to calculate the discrete Fourier transform of $\{f(k)\}$ requires N^2 complex operations (additions and multiplications), whereas to calculate the discrete Fourier transform of $\{f_1(k)\}$ or $\{f_2(k)\}$ requires only M^2 or $N^2/4$ complex operations. When we use Equation (4.19) to obtain $\{F(j)\}$ from $\{F_1(j)\}$ and $\{F_2(j)\}$, we require $N + 2(N^2/4)$ complex operations. In other words, we first require $2(N^2/4)$ operations to calculate the two Fourier transforms $\{F_1(j)\}$ and $\{F_2(j)\}$, and then we require the N additional operations prescribed by Equation (4.19). Thus we have reduced the number of operations from N^2 to $N + N^2/2$. For the smallest value of N (i.e., $N = 4$) this results in a factor of 0.75. When N is large, this factor approaches a factor of $\tfrac{1}{2}$.

Before proceeding, we make note of the important fact that when $N = 2$ we divide the second order sequence $\{f(k)\} = \{f(0), f(1)\}$ into two first order sequences $\{f_1(k)\} = \{f(0)\}$ and $\{f_2(k)\} = \{f(1)\}$. However, since a first order sequence is its own transform [i.e., $F_1(0) = f_1(0)$ and $F_2(0) = f_2(0)$], we do not require any complex multiplications or additions to obtain these transforms. Therefore, using Equation (4.19) for this case would require only $N = 2$ operations to obtain $F(j)$. For example, if $\{f(k)\} = \{1, 2\}$, then $\{f_1(k)\} = \{1\}$ and $\{f_2(k)\} = \{2\}$. Also $\{F_1(j)\} = \{f_1(k)\} = \{1\}$ and $\{F_2(j)\} = \{f_2(k)\} = \{2\}$. Now using Equation (4.19) we have:

$$F(0) = \tfrac{1}{2}\big(F_1(0) + F_2(0)W_2^0\big) = \tfrac{1}{2}(1 + 2) = \tfrac{3}{2},$$

$$F(1) = \tfrac{1}{2}\big(F_1(0) - F_2(0)W_2^0\big) = \tfrac{1}{2}(1 - 2) = -\tfrac{1}{2}.$$

Thus we obtain the result $\{f(j)\} = \{\tfrac{3}{2}, -\tfrac{1}{2}\}$ with only two complex operations. If Equation (4.1) were used, then four complex operations would have been required.

Now suppose that N is divisible by 4 or $M = N/2$ is divisible by 2. Then the subsequences $\{f_1(k)\}$ and $\{f_2(k)\}$ can be further subdivided into four $M/2$

order sequences as per Equation (4.15):

$$g_1(k) = f_1(2k),$$

$$g_2(k) = f_1(2k + 1),$$

$$h_1(k) = f_2(2k) \qquad k = 0, 1, \dots \frac{M}{2} - 1$$

$$h_2(k) = f_2(2k + 1).$$

Therefore, we can also use Equation (4.19) to obtain the Fourier transforms $\{F_1(j)\}$ and $\{F_2(j)\}$ with only $M + M^2/2$ complex operations and then use these results to obtain $\{F(j)\}$. A little thought reveals that this requires $N + 2(M + M^2/2) = 2N + N^2/4$ operations.

Thus when we subdivide a sequence twice ($N > 4$ and N divisible by 4) we reduce the number of operations from N^2 to $2N + N^2/4$. The $2N$ term is the result of applying Equation (4.19) (twice) whereas the $N^2/4$ term is the result of transforming the four reduced sequences. For the case when $N = 4$ we note that we completely reduce the sequence to four first order sequences that are their own transforms and, therefore, we do not need the additional $N^2/4$ transform operations. The formula then becomes $2N$. The smallest value of N that does not result in complete reduction of the sequence is 8. For this case we have a reduction factor of $\frac{1}{2}$, whereas for large N the factor approaches $\frac{1}{4}$. Continuing in this way we can show that if N is divisible by 2^p (p is a positive integer), then the number of operations required to compute the discrete Fourier transform of the Nth order sequence $\{f(k)\}$ by repeated subdivision is

$$pN + \frac{N^2}{2^p}.$$

Again for complete reduction (i.e., $N = 2^p$) the $N^2/2^p$ term is not required and we obtain pN for the number of operations required. This results in a reduction factor of

(4.20)
$$\frac{pN}{N^2} = \frac{p}{N} = \frac{\log_2 N}{N}.$$

The essence of the Cooley–Tukey algorithm is to choose sequences with $N = 2^p$ and go to complete reduction. Although most sequences do not have such a convenient number of terms, we can always artificially add zeros to the end of the sequence to reach such a value. This extra number of terms in the sequence is more than compensated for by the tremendous savings afforded by using the Cooley–Tukey algorithm.

MIXED RADIX FAST TRANSFORMS

In the previous section we considered an algorithm for computing the discrete Fourier transform of an Nth order sequence that contained an even number of terms; that is, N was divisible by 2. We now parallel that development for sequences in which the number of terms N is divisible by 3; that is, $L = N/3$ is an integer. We begin by assuming that we have an Nth order sequence $\{f(k)\}$ with discrete Fourier transform $\{F(j)\}$. Also, we assume that N is divisible by 3 and thus it makes sense to consider the three new sequences

$$f_1(k) = f(3k),$$

$$(4.21) \quad f_2(k) = f(3k + 1), \qquad k = 0,\dots, L - 1, \text{ where } L = N/3.$$

$$f_3(k) = f(3k + 2),$$

As an example, using Equation (4.21), the ninth order sequence $\{f(k)\} = \{1, 3, 7, 5, 9, 2, 4, 8, 6\}$ produces the three following third order subsequences: $\{f_1(k)\} = \{1, 5, 4\}$, $\{f_2(k)\} = \{3, 9, 8\}$, and $\{f_3(k)\} = \{7, 2, 6\}$.

It can be easily shown that the three subsequences defined by Equation (4.21) are all periodic with periodicity $L = N/3$. Also, since L is an integer we can use Equation (4.1) to determine their discrete Fourier transforms:

$$F_1(j) = \frac{1}{L} \sum_{k=0}^{L-1} f_1(k) W_L^{-kj},$$

$$(4.22) \qquad F_2(j) = \frac{1}{L} \sum_{k=0}^{L-1} f_2(k) W_L^{-kj}, \qquad j = 0,\dots, L - 1.$$

$$F_3(j) = \frac{1}{L} \sum_{k=0}^{L-1} f_3(k) W_L^{-kj}.$$

Theorem 4.2 guarantees that $\{F_1(j)\}$, $\{F_2(j)\}$, and $\{F_3(j)\}$ are all periodic with periodicity L. Let us now consider the Fourier transform of the Nth order sequence $\{f(k)\}$:

$$F(j) = \frac{1}{N} \sum_{k=0}^{N-1} f(k) W_N^{-kj}.$$

Splitting the summation we can rewrite the preceding equation as

$$F(j) = \frac{1}{N} \sum_{k=0}^{L-1} f(3k)W_N^{-3kj} + \frac{1}{N} \sum_{k=0}^{L-1} f(3k+1)W_N^{-(3k+1)j}$$

(4.23)

$$+ \frac{1}{N} \sum_{k=0}^{L-1} f(3k+2)W_N^{-(3k+2)j}.$$

However,

$$W_N^{-3kj} = W_L^{-kj},$$

(4.24)
$$W_N^{-(3k+1)j} = W_L^{-kj}W_N^{-j},$$

$$W_N^{-(3k+2)j} = W_L^{-kj}W_N^{-2j},$$

and, therefore, Equation (4.23) becomes

(4.25)

$$F(j) = \tfrac{1}{3}\big[F_1(j) + W_N^{-j}F_2(j) + W_N^{-2j}F_3(j)\big], \qquad j = 0,\ldots, N-1.$$

However, since $\{F_1(j)\}$, $\{F_2(j)\}$, and $\{F_3(j)\}$ are periodic (with periodicity L), we have

$$F(j) = \tfrac{1}{3}\big[F_1(j) + W_N^{-j}F_2(j) + W_N^{-2j}F_3(j)\big],$$

(4.26) $\quad F(j+L) = \tfrac{1}{3}\big[F_1(j) + e^{-2\pi i/3}W_N^{-j}F_2(j) + e^{-4\pi i/3}W_N^{-2j}F_3(j)\big],$

$$F(j+2L) = \tfrac{1}{3}\big[F_1(j) + e^{-4\pi i/3}W_N^{-j}F_2(j) + e^{-2\pi i/3}W_N^{-2j}F_3(j)\big],$$

$$j = 0,\ldots, L-1.$$

To calculate the discrete Fourier transform of $\{f(k)\}$ requires N^2 complex operations (additions and multiplications), whereas to calculate the discrete Fourier transform of the subsequences $\{f_1(k)\}$, $\{f_2(k)\}$, or $\{f_3(k)\}$ requires only L^2 or $N^2/9$ complex operations. When we use Equations (4.26) to obtain $\{F(j)\}$ from $\{F_1(j)\}$, $\{F_2(j)\}$, and $\{F_3(j)\}$, we require $3N + N^2/3$ complex operations. That is, we first require $3(N^2/9)$ operations to calculate the three transforms $\{F_1(j)\}$, $\{F_2(j)\}$, and $\{F_3(j)\}$, and then we require $3N$ additional operations as spelled out by Equations (4.26). Thus we have reduced the number of operations from N^2 to $3N + N^2/3$. For $N = 9$ (the smallest value of N that does not result in complete reduction), this results in a factor of $\tfrac{2}{3}$. When N is large this factor approaches $\tfrac{1}{3}$.

It can be shown that if we can subdivide the sequence twice (i.e., N is divisible by 9), then the number of operations required to calculate its discrete Fourier transform is $2(3N) + N^2/9$. For large values of N, the savings factor approaches $\frac{1}{9}$. In the general case, when the sequence can be subdivided q times (N divisible by 3^q), it can also be shown that the repeated use of Equation (4.26) allows us to calculate the discrete Fourier transform of the sequence with only $q(3N) + N^2/3^q$ operations. For complete reduction (i.e., $N = 3^q$) the $N^2/3^q$ term is not required and we only require $q(3N)$ operations. This results in a reduction factor of

$$(4.27) \qquad \frac{q(3N)}{N^2} = \frac{3q}{N} = \frac{3\log_3 N}{N}.$$

It is interesting to note that if we do not take a sequence to complete reduction, then the radix 3 method affords more computational savings than does the radix 2. However, when we do go to complete reduction, the radix 2 method is superior. For example, if $N = 36$, then we can reduce this sequence twice using either the radix 2 or radix 3 method (N is divisible by both 9 and 4). The radix 2 method requires $2N + N^2/4 = 396$ operations whereas the radix 3 method only requires $2(3N + N^2/9) = 360$ operations. Suppose that we wish to determine the discrete Fourier transform of a sequence with 700 terms. Since 700 is neither a power of two or three we must artificially add zeros to this sequence. For radix 3, the next greatest integer that is a power of 3 is $729 = 3^6$ and thus we must add 29 zeros. To go to complete reduction now requires $6[3(729)] = 13122$ operations. For the radix 2 method we must add 324 zeros to obtain $N = 1024 = 2^{10}$. To calculate the discrete Fourier transform of this sequence requires $10(1024) = 10240$ operations which is a factor of 0.78 less than the radix 3 method.

In an analogous way it is possible to generate radix 5, radix 7, and higher radix methods. It is also possible to combine these various methods into one algorithm. That is, suppose

$$N = 2^q 3^q 5^r.$$

To digitally obtain the discrete Fourier transform of this Nth order sequence we would first subdivide the sequence r times using the radix 5 method, then q times using the radix 3 method, and finally p times using the radix 2 method. These methods can easily be programmed for use on a digital computer. Most computer libraries have at least one such fast Fourier transform code available.

SUMMARY

In this chapter we presented the discrete Fourier transform, which is an operation that maps an Nth order sequence to another Nth order sequence.

We then presented several theorems that described various properties of this transform. These theorems are analogous to those presented in Chapters 2 and 3 for the Fourier series and Fourier transform. Finally, we presented techniques and algorithms that permit the rapid and efficient calculation of the discrete Fourier transform of a sequence.

BIBLIOGRAPHY

Cooley, J. W., P. A. W. Lewis, and P. D. Welsh, "The Finite Fourier Transform," *IEEE Trans. Audio-Electroacoustics*, **AV-17**, No. 2, June 1969.

Freeman, H., *Discrete Time Systems, An Introduction to the Theory*, Wiley, New York, 1965.

Weaver, H. J., *An Introduction to Discrete Fourier Analysis*, University of California, Lawrence Livermore Laboratory, Report No. UCRL 73113, Livermore, Calif., 1971.

Cooley, J. W., and J. W. Tukey, "An Algorithm for the Machine Calculation of Complex Fourier Series," *Math. Computations*, **19**, April 1965.

Singleton, R. C., "A Short Bibliography on the Fast Fourier Transform," *IEEE Trans. on Audio and Electroacoustics*, **AU-17**, No. 2, June 1966.

G-AE Subcommittee on Measurement Concepts, "What is the Fast Fourier Transform," *IEEE Trans. on Audio and Electroacoustics*, **AU-15**, No. 2, June 1967.

CHAPTER

5

FOURIER ANALYSIS VIA
A DIGITAL COMPUTER

In the previous chapters we presented a discussion of three different transformations. In Chapter 2 we examined the Fourier series which may be considered a transformation that maps an analytical function to a sequence of (Fourier series) coefficients. In Chapter 3 we saw that the Fourier transform was a mapping that took a function $f(t)$ to another function $F(w)$ (its Fourier transform). Finally in Chapter 4 we presented the discrete Fourier transform that maps a sequence $\{f(k)\}$ to its transform sequence $\{F(j)\}$. These transforms are illustrated schematically in Figure 5.1.

Both the Fourier series and Fourier transform require the analytical determination of an integral equation. Depending upon the complexity of the function involved, these integrals can be rather difficult, if not impossible, to evaluate. However, the discrete Fourier transform deals with bounded sequences and requires only straightforward addition and multiplication of terms. Furthermore, in Chapter 4 we presented several properties and algorithms (in particular the FFT algorithm) that permit us to calculate the discrete Fourier transform in a very rapid and efficient manner.

If we compare the basic definition and property theorems of the Fourier series, the Fourier transform, and the discrete Fourier transform, we note rather obvious similarities among them. In this chapter we take advantage of these similarities and show how the relatively simple discrete Fourier transform

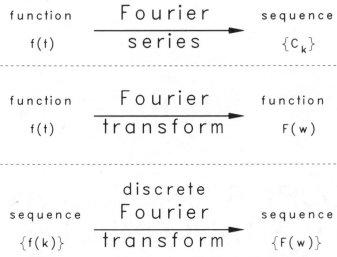

FIGURE 5.1.   Schematic illustration of the Fourier transformations.

can be used to obtain both the Fourier transform and Fourier series coefficients of a function. The basic approach that we use is to convert the function to a sequence (by sampling), calculate the discrete Fourier transform of this sequence, and from this obtain either the Fourier series or Fourier transform of the original function.

SAMPLING A FUNCTION

In the real world we are often interested in recording and analyzing signals from some phenomenon or physical event. Although these signals can rarely, if ever, be characterized by an analytical expression, they are almost always well enough behaved so that their Fourier transforms exist. Because these functions do not possess a mathematical, or analytical, expression they must first be digitized and then analyzed by means of numerical algorithms on a digital computer.

When we digitize, or sample, a function $f(x)$ we, in effect, convert that function to a sequence $\{\tilde{f}(k)\}$. We do so by choosing values of that function at discrete locations of x. We assume here that these discrete locations are evenly spaced in x with the distance between any two samples being Δx. The $\tilde{f}(k)$ term of the sequence is equal to the value of the function $f(x)$ at $x = x_0 + k\Delta x$ [x_0 is the location of the first ($k = 0$) sample]. Naturally, we want this sampled sequence to properly represent the function. We consider the sequence $\{\tilde{f}(k)\}$ to be an adequate representation of the function $f(x)$ if we can recover that function exactly from the sequence. That is to say, if we can somehow interpolate between the sequence terms $\tilde{f}(k)$ to return the function $f(x)$. This concept is discussed in greater detail in the next sections; however, here we heuristically discuss some of the problems encountered when we sample a function. To illustrate these problems let us assume that we wish to sample the function shown in Figure 5.2. To obtain an accurate representation of this function, the sampling size (or sampling rate) Δx must be small in comparison

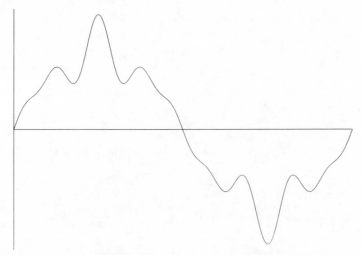

FIGURE 5.2. Function to be sampled.

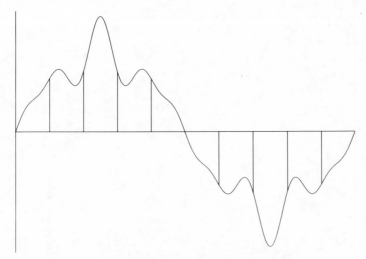

FIGURE 5.3. Example of aliasing.

with the "detail" of the function. Just how small Δx must be is rigorously spelled out by the celebrated sampling theorem presented in the next section. A rough rule of thumb to determine how accurately the sampled sequence represents the function is to construct a smooth curve between the sampled values; the better the representation—the closer this curve will lie to the original function Clearly, the sampling scheme shown in Figure 5.3 is a poor representation of the original function because Δx is too large. When this is the situation we say that the data has been *aliased*.

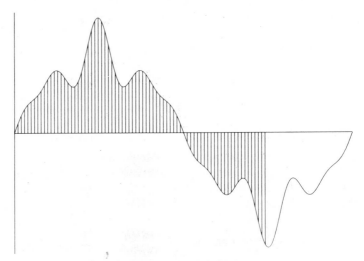

FIGURE 5.4.   Example of windowing.

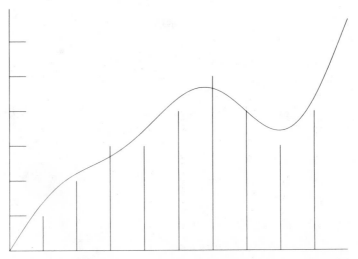

FIGURE 5.5.   Example of quantization error.

The second common difficulty encountered in sampling a function is that of *windowing*. An example of this is shown in Figure 5.4. In this case our sampling rate Δx appears to be adequate; however, instead of sampling the entire function we have sampled only a portion of it. Figuratively, we have placed a window over the function.

Quantization, the third common difficulty, is caused because we cannot digitize a function with infinitesimal accuracy but instead with only discrete levels. A "blowup" of a portion of our function is shown in Figure 5.5. In this case our digitizing machine can only put out the discrete values indicated by the hash marks on the vertical axis. Thus the actual value of the function (at the sampled points) is "rounded off" to the nearest digital level.

THE SAMPLING THEOREM

In this section we present the sampling theorem which theoretically answers the question of how small our sample size Δx must be in order to prevent aliasing of the data. The sampling theorem also describes an interpolation rule by which we can uniquely recover a function from its sampled sequence. We begin by considering a function $f(x)$ whose Fourier transform is equal to zero outside the interval $[-\Omega, \Omega]$:

$$F(w) = 0 \quad \text{if } |w| \geqslant \Omega > 0.$$

Such functions $f(x)$ are called *band-limited* with bandwidth 2Ω. Because $f(x)$ is band-limited, we are able to generate a protracted version of its Fourier transform $F(w)$, call it $G(w)$ (see Chapter 2). In this chapter we assume that $G(w)$ satisfies the Dirichlet conditions. This fact permits us to use Equations (2.9) and (2.12) to represent it as a Fourier series:

$$(5.1) \qquad G(w) = \sum_{k=-\infty}^{\infty} C_k e^{2\pi i k w/2\Omega},$$

where

$$(5.2) \quad C_k = \frac{1}{2\Omega} \int_{-\Omega}^{\Omega} G(w) e^{-2\pi i k w/2\Omega} \, dw = \frac{1}{2\Omega} \int_{-\Omega}^{\Omega} F(w) e^{-2\pi i k w/2\Omega} \, dw.$$

Now let us independently use Equation (3.2) and consider the expression for the inverse Fourier transform of $F(w)$:

$$f(x) = \int_{-\infty}^{\infty} F(w) e^{2\pi i w x} \, dw.$$

Substitution of the discrete values $-k/2\Omega$ for x in the preceding equation yields

$$(5.3) \quad f\left(\frac{-k}{2\Omega}\right) = \int_{-\infty}^{\infty} F(w)e^{-2\pi ikw/2\Omega}\,dw = \int_{-\Omega}^{\Omega} F(w)e^{-2\pi ikw/2\Omega}\,dw.$$

Comparing Equations (5.2) and (5.3) we see

$$(5.4) \qquad\qquad 2\Omega C_k = f\left(\frac{-k}{2\Omega}\right) \quad \text{for all (integer) } k.$$

Thus the Fourier series coefficients of the function $F(w)$ are isomorphic to the value of its (inverse) Fourier transform $f(x)$ at the discrete locations $x = -k/2\Omega$. Substitution of Equation (5.4) into Equation (5.1) results in

$$G(w) = \sum_{k=-\infty}^{\infty} \frac{1}{2\Omega}f\left(\frac{-k}{2\Omega}\right)e^{2\pi ikw/2\Omega}.$$

However, over the interval $[-\Omega, \Omega]$ we have $G(w) = F(w)$ and, therefore,

$$F(w) = \sum_{k=-\infty}^{\infty} \frac{1}{2\Omega}f\left(\frac{-k}{2\Omega}\right)e^{2\pi ikw/2\Omega}, \qquad w \in [-\Omega, \Omega].$$

Now taking the inverse Fourier transform of both sides of the preceding equation we find

$$f(x) = \int_{-\infty}^{\infty} F(w)e^{2\pi iwx}\,dw, = \int_{-\Omega}^{\Omega} F(w)e^{2\pi iwx}\,dw,$$

$$= \int_{-\Omega}^{\Omega}\left[\sum_{k=-\infty}^{\infty} \frac{1}{2\Omega}f\left(\frac{-k}{2\Omega}\right)e^{2\pi ikw/2\Omega}\right]e^{2\pi iwx}\,dw.$$

Assuming that we can interchange the summation and integral signs (this is almost always valid for real world functions) and a little algebra we arrive at:

$$f(x) = \sum_{k=-\infty}^{\infty} f\left(\frac{-k}{2\Omega}\right)\frac{1}{2\Omega}\int_{-\Omega}^{\Omega} e^{2\pi iw(x+k/2\Omega)}\,dw,$$

$$f(x) = \sum_{k=-\infty}^{\infty} f\left(\frac{-k}{2\Omega}\right)\frac{\sin 2\pi\Omega[k\Delta x + x]}{2\pi\Omega[k\Delta x + x]}, \quad \text{where } \Delta x = \frac{1}{2\Omega}.$$

If we now let $\tilde{f}(k) = f(k\Delta x)$ and substitute $-k$ for k, we find,

$$(5.5) \qquad\qquad f(x) = \sum_{k=-\infty}^{\infty} \tilde{f}(k)\frac{\sin(2\pi\Omega[x - k\Delta x])}{2\pi\Omega[x - k\Delta x]}.$$

We summarize the previous development in the following theorem.

Theorem 5.1 (Sampling Theorem). If the function $f(x)$ is band-limited with bandwidth 2Ω:

$$F(w) = 0 \quad \text{for } |w| \geqslant \Omega > 0,$$

then $f(x)$ is uniquely determined by a knowledge of its values at uniformly spaced intervals Δx apart ($\Delta x = 1/2\Omega$). Specifically, we have

$$f(x) = \sum_{k=-\infty}^{\infty} f(k\Delta x) \frac{\sin(2\pi\Omega[x - k\Delta x])}{2\pi\Omega[x - k\Delta x]}.$$

Let us take a closer look at this sampling theorem to see just what it can do for us. First of all, it tells us how often we must sample a function in order to be able to uniquely recover the function from its sampled sequence. In other words, for this band-limited function, the highest frequency component present is Ω and the sampling theorem requires that we must have a sampling rate of at least $1/2\Omega$. Physically, this means we must have at least two samples per cycle of the highest frequency component present. This sampling rate ($\Delta x = 1/2\Omega$) is often called the *Nyquist rate* and the sequence obtained using this rate, the *Nyquist samples*.

In addition to this information, the sampling theorem also supplies us with an interpolation formula with which to recover the function from its Nyquist samples. Basically, we construct a sinc function about each sampled value. The sum of all these "weighted" sinc functions is then exactly equal to the original function. For example, in Figure 5.6 we show an expanded view of this

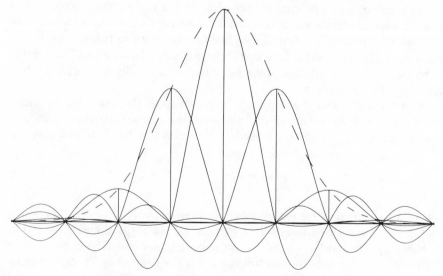

FIGURE 5.6. Sampling theorem interpolation formula.

interpolation procedure for a portion of the function containing five Nyquist samples. The dashed line is the sum of the individual sinc functions. *Note:* In this figure, for clarity, the "tails" of the sinc functions from the other sampled values are not shown. The sum, however (dashed line), was constructed using them.

So much for the good news, now the the bad. The sampling theorem applies to the ideal case in which we can sample the function with infinite accuracy and in which there are no quantization or windowing errors. It also assumes that the function is sampled from $-\infty$ to ∞, which obviously presents some problems in actual practice. The sampling theorem, therefore, must be considered as an ideal model meant to give insight into the problem of sampling a function. It is to be used as a guide and not as a sacrosanct or unbreakable rule.

It turns our that in actual practice about 10 samples per cycle of the highest frequency is usually an acceptable sampling rate. To determine the sampling rate we must know the highest frequency present in the function or, in other words, we must know the Fourier transform of the function. However, the motivation for digitizing a function in the first place is often to calculate its Fourier transform (on a computer). The way out of this apparent dilemma is to use common sense and a little intuition. That is, to sample a function when we don't know its bandwidth *a priori*, we should choose the sampling rate such that a smooth curve drawn through the samples will "strongly resemble" the original function. If we follow this general rule of thumb, then for most all practical applications the function will be sampled fine enough. We also point out here that the interpolation rule presented by the sampling theorem is very tedious and expensive (in terms of computer time). Again, if we use the real world sampling rate (10 samples per cycle of the highest frequency), then a smooth curve (often a series of quadratics or even linear equations) drawn through the values will return the original function with sufficient accuracy. It must be clearly understood here that the term "sufficient accuracy" is a rather nebulous one. Each situation must be judged individually and accuracy balanced against expediency.

As previously noted, the sampling theorem requires that our samples extend from $-\infty$ to ∞; in other words, our sampled sequence contains an infinite number of terms. Let us now consider a function that has bounded support:

$$f(x) = 0, \qquad |x| \geqslant X,$$

where X is some positive real number. In Fourier analysis such a function is called *time-limited* or *space-limited* (depending upon the dimensions of the variable x). Obviously, we need only sample time-limited functions over a finite domain. For time-limited functions it is tempting to apply the sampling theorem and write the interpolation formula (over a finite domain) as

$$f(x) = \sum_{k=-M}^{M} f(k\Delta x)\frac{\sin(2\pi\Omega[x - k\Delta x])}{2\pi\Omega[x - k\Delta x]}.$$

However, this is not permitted because the sampling theorem requires that the function be also band-limited and, in the strict mathematical sense, this is a violation of the uncertainty principle. We once again look to the real world for guidance and note that there are many functions that are "almost" band-limited and/or "almost" time-limited. More formally, we say a function is *almost time-limited* if and only if given any $\varepsilon_T > 0$ there exists a positive real number X (called the time limit) such that

$$\int_{-\infty}^{-x} |f(x)|\, dx < \varepsilon_T, \quad \text{and} \quad \int_{x}^{\infty} |f(x)|\, dx < \varepsilon_T.$$

Similarly, a function $f(x)$ with Fourier transform $F(w)$ is called *almost band-limited* if and only if given any $\varepsilon_B > 0$, there exists a positive real number Ω (called the band limit) such that

$$\int_{-\infty}^{-\Omega} |F(w)|\, dw < \varepsilon_B \quad \text{and} \quad \int_{\Omega}^{\infty} |F(w)|\, dw < \varepsilon_B.$$

An example of a function that is both almost time-limited and almost band-limited is the Gaussian function

$$f(x) = e^{-ax^2}.$$

All real signals are almost band-limited and most of them are almost time-limited. The band and time limits are determined, to a large degree, by the resolution, sensitivity, and/or dynamic range of the detection instruments.

We now present (without proof) a more useful version of the sampling theorem.

Theorem 5.2 (Real World Sampling Theorem). If the function $f(x)$ is almost band-limited, with bandwidth 2Ω, and almost time-limited, with time width $2X$, then $f(x)$ can be recovered from its sampled sequence to any desired accuracy. That is, given any $\varepsilon > 0$ we can always choose X and Ω such that

$$f(x) = \sum_{k=-M}^{M} f(k\Delta x)\frac{\sin(2\pi 5\Omega[x - k\Delta x])}{2\pi 5\Omega[x - k\Delta x]} + R_\varepsilon(x).$$

where

$$|R_\varepsilon(x)| < \varepsilon,$$

$$M\Delta x > X,$$

$$\Delta x < \frac{1}{10\Omega}.$$

Note that in this theorem we used the more realistic sampling rate of 10 samples per highest frequency.

COMPUTER-CALCULATED FOURIER TRANSFORMS

In this section we present a method by which we can obtain the Fourier transform of a function from the discrete Fourier transform of the sequence obtained by sampling the function. We begin by considering an almost time-limited function such as the one shown in Figure 5.7. We assume that the time limits are 0 and X. We also assume that this function is almost band-limited by the domain $[-\Omega, \Omega]$. We next sample this function with the (real world) sampling rate of $\Delta x = 1/10\Omega$ and choose the number of samples N in such a

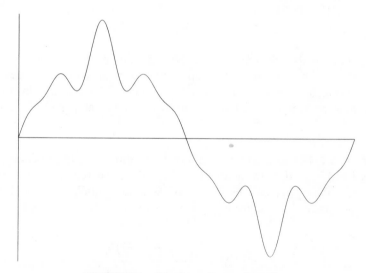

FIGURE 5.7.   Function to be transformed.

way that we avoid any windowing problems; that is, $N\Delta x \geqslant 10X^\dagger$. In this way we have produced a sequence $\{f(k\Delta x)\}$ or simply $\{f(k)\}$. Furthermore, Theorem 5.2 guarantees that we can recover the original function $f(x)$ from this sequence as per the formula

(5.6) $$f(x) = \sum_{k=0}^{N-1} f(k\Delta x)\text{sinc}(2\pi 5\Omega[x - k\Delta x]).$$

Note that in the preceding equation we have assumed that the remainder term $R_\varepsilon(x)$ is negligible and consequently we have dropped it. Taking the Fourier transform of both sides of Equation (5.6) we find

$$F(w) = \mathcal{F}\left[\sum_{k=0}^{N-1} f(k\Delta x)\text{sinc}(2\pi 5\Omega[x - k\Delta x]) \right],$$

or

(5.7) $$F(w) = \sum_{k=0}^{N-1} f(k\Delta x)\mathcal{F}\left[\text{sinc}(2\pi 5\Omega[x - k\Delta x])\right].$$

In Chapter 3 we showed that the sinc function and pulse function are Fourier transform pairs. Using this result and the shifting Theorem 3.2 with $a = k\Delta x$ we find

$$\mathcal{F}\left[\text{sinc}(2\pi 5\Omega[x - k\Delta x])\right] = \frac{1}{10\Omega} P_{5\Omega}(w)e^{-2\pi iwk\Delta x}.$$

Substituting this into Equation (5.7) we obtain

$$F(w) = \frac{1}{10\Omega} \sum_{k=0}^{N-1} f(k\Delta x) P_{5\Omega}(w)e^{-2\pi iwk\Delta x}.$$

We now have an expression for $F(w)$, the Fourier transform of $f(x)$, in the form of a (finite) series expansion weighted by the terms of the sampled sequence of $f(x)$. Further examination of the function $F(w)$ reveals that it may be considered a time-limited function over the domain $[-\Omega, \Omega]$. Also, since its Fourier transform is $f(-x)$ (see Theorem 3.6), it is also band-limited over the domain $[-X, 0]$. Let us now form the Nth order sequence $\{F(j)\}$ by sampling $F(w)$ [as given by Equation (5.7)] over the domain $[-5\Omega, 5\Omega]$. We are using N samples and this results in a sampling rate of $\Delta w = 10\Omega/N$. Since the domain $[-5\Omega, 5\Omega]$ is five times larger than the time width of $F(w)$, there are obviously no windowing problems. Furthermore, this sampling rate satisfies Theorem 5.2:

$$\Delta w = \frac{10\Omega}{N} = \frac{1}{N\Delta x} = \frac{1}{10X}.$$

---

†Actually this domain is about 10 times greater than we need to avoid windowing the function. However, as we continue with the presentation the reason becomes clear.

Consequently, aliasing will not be a problem. We can now write

$$F(j\Delta w) = \frac{1}{10\Omega} \sum_{k=0}^{N-1} f(k\Delta x) P_{5\Omega}(w) e^{2\pi i k j\Delta w \Delta x}.$$

However, we note that

$$P_{5\Omega}(w) = 1 \quad \text{over the sampled domain,}$$

$$\Delta w \Delta x = \frac{10\Omega}{N} \frac{1}{10\Omega} = \frac{1}{N},$$

$$\frac{1}{10\Omega} = \frac{N}{10\Omega} \frac{1}{N} = N\Delta x \left(\frac{1}{N} \right).$$

Thus the previous equation becomes

$$F(j\Delta w) = N\Delta x \frac{1}{N} \sum_{k=0}^{N-1} f(k\Delta x) e^{-2\pi i k j/N},$$

or

(5.8) $$\{F(j)\} = N\Delta x \overline{\overline{\mathcal{F}}} [\{f(k)\}].$$

Thus we see that the sequence formed by sampling the Fourier transform $F(w)$ is directly proportional to the discrete Fourier transform of the sequence formed by sampling $f(x)$. Furthermore, the way our sampling scheme was chosen, we know that $F(w)$ and $f(x)$ can be uniquely recovered from these sampled sequences.

In the preceding development it was assumed that the function $f(x)$ was time limited by 0 and on the left:

$$f(x) = 0 \quad \text{for } x < 0.$$

If this is not the case, then we simply form a new function $g(x)$ by shifting $f(x)$ to the right by some amount A so that $g(x)$ will be time-limited on the left by 0. That is,

$$g(x) = f(x - A) \quad \text{where } A \text{ is chosen such that}$$

$$g(x) = 0 \qquad \text{for all } x < 0.$$

Now this development can be applied to this new function $g(x)$ and we can obtain its Fourier transform $G(w)$. The first shifting Theorem 3.2 tells us that $G(w)$ and $F(w)$ are related by

$$G(w) = F(w) e^{-2\pi i w A}.$$

To recover $F(w)$ we write

$$F(w) = G(w)e^{2\pi iwA}.$$

We now summarize the overall procedure.

Assume that $f(x)$ is an almost time-limited function over the domain $[-A, X]$. Let us also assume that it is almost band-limited over the domain $[-\Omega, \Omega]$. Then to digitally obtain the Fourier transform of this function we proceed as follows:

1. *If necessary ($A \neq 0$), form the new function $g(x)$ by shifting $f(x)$ to the right by an amount A:*

$$g(x) = f(x - A).$$

2. *Sample the function $g(x)$ with sampling rate $\Delta x = 1/10\Omega$ and choose the number of samples N such that*

$$N\Delta x > 10(X + A).$$

3. *Calculate the discrete Fourier transform of this sampled sequence and multiply the resulting sequence by $N\Delta x$ to obtain the sequence $\{G(j\Delta w)\}$.*

4. *Use Theorem 4.3 to obtain values for the negative indices j that represent values for the negative frequencies $-j\Delta w$ ($\Delta w = 1/N\Delta x$).*

5. *Recover $G(w)$ from $\{G(j\Delta w)\}$ as per the sampling Theorem 5.2 (or by simply constructing a smooth curve between the sampled values).*

6. *If necessary ($A \neq 0$), recover $F(w)$ from $G(w)$ as per the formula*

$$F(w) = G(w)e^{2\pi iAw}.$$

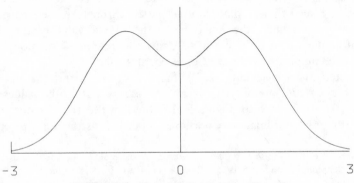

FIGURE 5.8. Function $f(x)$ to be transformed.

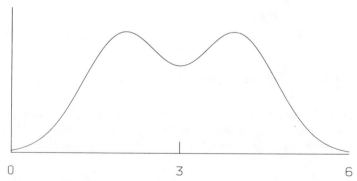

FIGURE 5.9. The shifted function $g(x) = f(x - 3)$.

NOTE: If $A = 0$, then $F(w) = G(w)$.

As an illustration of this procedure we digitally determine the Fourier transform of the (real and even) function shown in Figure 5.8 which ranges over the domain $[-3, 3]$. We first note that this function is not time-limited on the left by 0 and, therefore, we must shift it to the right by an amount $A = 3$ as per the instructions in step 1. That is to say, we form a new function $g(x)$ given by $g(x) = f(x - 3)$. This is shown in Figure 5.9.

Step 2 requires that we sample this function as per Theorem 5.2 (i.e., $\Delta x = 1/10\Omega$) and choose the number of samples such that

$$(5.9) \qquad N\Delta x \geqslant 10(X + A) = 10(3 + 3) = 60.$$

As is most often the case, we don't know the value of the band limit Ω and, therefore, we have to guess at a value of Δx by an examination of the time domain function $g(x)$. As discussed earlier in the chapter, we choose a sampling rate such that a smooth curve passed through the sequence values will "strongly resemble" the function. In the case a value of $\Delta x = 0.2$ is sufficient as illustrated in Figure 5.10. Using this rate and Equation (5.9) we determine the number of samples N to be equal to 300.[†]

Next (step 3) we calculate $\{G(j)\}$, the discrete Fourier transform of the sequence $\{g(k)\}$ and multiply the terms by $N\Delta x$ to obtain the values of the sequence $\{G(j\Delta w)\}$ shown in Figure 5.11. Note that this transform is complex valued with a real and imaginary portion. As per the instructions in step 4, we

---

[†]For the sake of clarity we use $N = 300$. However, in actual practice we would adjust the time limit or the sampling rate so that N is equal to a power of two. In this way the power of the FFT algorithm can be employed to calculate the discrete Fourier transform.

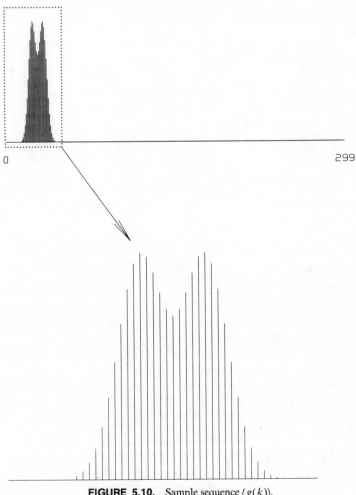

FIGURE 5.10. Sample sequence $\langle g(k) \rangle$.

use Theorem 4.3 to obtain values for the negative indices:

$$G(-j) = G(N-j).$$

The results of this move are shown in Figure 5.12.

In step 5 we recover $G(w)$, the transform of $g(x)$, by passing a smooth curve between the values of the sequence $\langle G(j) \rangle$. This is shown in Figure 5.13. Note that the frequency scale is generated as per the equation $\Delta w = 1/(N\Delta x) = 1/60$. Finally, the desired result $F(w)$ is obtained, as per step 6, by

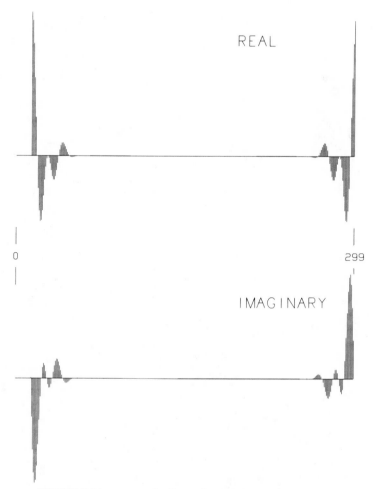

FIGURE 5.11. $\langle G(j)\rangle$, the discrete Fourier transform of $\langle g(k)\rangle$.

multiplying $G(w)$ by

$$e^{2\pi i wA} = e^{2\pi i w3} = \cos 6\pi w + i \sin 6\pi w.$$

This is shown in Figure 5.14. Note that this transform is real and even, as it certainly must be since $f(x)$ was real and even (see Theorem 3.14i).

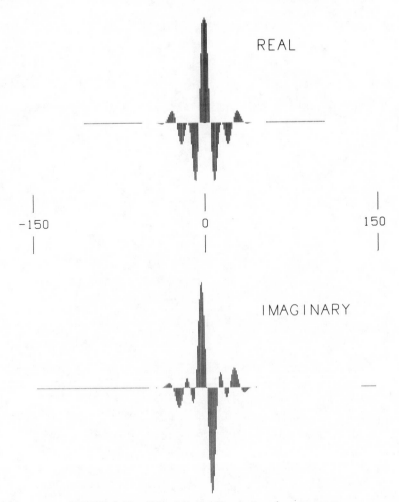

REAL

−150   0   150

IMAGINARY

FIGURE 5.12.   Shifted discrete Fourier transform sequence.

COMPUTER-GENERATED FOURIER SERIES

In this section we extend several of the results of the previous section and present a method that permits the calculation of the Fourier series coefficients of a function on a digital computer. Just as in the previous section, we accomplish this by sampling the function and obtaining the desired coefficients from the terms of the discrete Fourier transform of the sampled sequence.

We begin by considering a function $g(t)$ that satisfies the Dirichlet conditions with period equal to T. We now define a new function $f(t)$ equal to $g(t)$

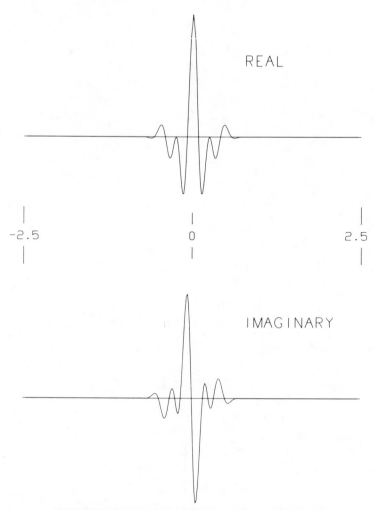

FIGURE 5.13. Transform $G(w)$ recovered from $\{G(j\Delta w)\}$.

over the domain $[-T/2, T/2]$ and equal to zero elsewhere:

$$f(t) = g(t) \qquad t \in \left[-\frac{T}{2}, \frac{T}{2}\right],$$

$$f(t) = 0, \qquad |t| > \frac{T}{2}.$$

First we write the Fourier series expression for $g(t)$:

$$g(t) = \sum_{k=-\infty}^{\infty} C_k e^{2\pi i k t/T},$$

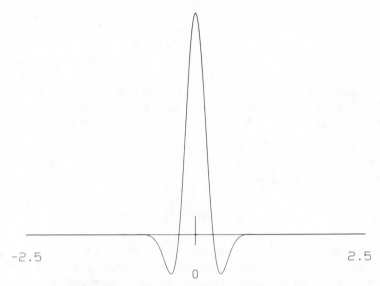

FIGURE 5.14.   Digitally obtained Fourier transform $F(w)$.

where

(5.10) $$C_k = \frac{1}{T}\int_{-T/2}^{T/2} g(t)e^{-2\pi ikt/T}\,dt = \frac{1}{T}\int_{-T/2}^{T/2} f(t)e^{-2\pi ikt/T}\,dt.$$

Now we write the Fourier transform expression for the function $f(t)$,

$$F(w) = \int_{-\infty}^{\infty} f(t)e^{-2\pi iwt}\,dt = \int_{-T/2}^{T/2} f(t)e^{-2\pi iwt}\,dt.$$

If, in the preceding equation we let $w = k/T$, we obtain

(5.11) $$F\left(\frac{k}{T}\right) = \int_{-T/2}^{T/2} f(t)e^{-2\pi ikt/T}\,dt.$$

Comparing Equations (5.10) and (5.11) we find

(5.12) $$C_k = \frac{1}{T}F\left(\frac{k}{T}\right).$$

Thus we see that the Fourier series coefficients of a function may be obtained from the Fourier transform of that function at equally spaced increments $w = k/T$. As an example, let us use this technique to calculate the Fourier series coefficients of the periodic function $g(t)$ defined over the basic interval

$[-1, 1]$ as

$$p(t) = \begin{cases} 1, & t \in [-\tfrac{1}{2}, \tfrac{1}{2}], \\ 0, & \text{otherwise.} \end{cases}$$

First we define $f(t)$ to be equal to $g(t)$ over the interval $[-1, 1]$ and 0 otherwise. The Fourier transform of this function is obtained by using Equation (3.11) with $a = \tfrac{1}{2}$:

$$F(w) = \text{sinc } \pi w.$$

Therefore, Equation (5.12) implies

$$C_k = \frac{1}{2} \text{sinc} \frac{\pi k}{2}.$$

This result may be verified by using Equation (3.12) with $b = \tfrac{1}{4}$ and $T = 2$.

The digital calculation of the Fourier series coefficients is now a straightforward application of Equation (5.12) and the results of the previous section. That is, consider the periodic function $g(t)$ with period T. We denote the function over its basic period of $[-T/2, T/2]$ as $f(t)$ and the transform of $f(t)$ as $F(w)$. We first sample $f(t)$ by selecting the sampling rate Δx (as per Theorem 5.2) and the number of samples N such that

(5.13) $N\Delta x = T.$

We denote the resulting sequence as $\{f(k)\}$ and its discrete Fourier transform as $\{F(j)\}$. Equation (5.8) tells us that

$$\{F(j\Delta w)\} = N\Delta x \overline{\overline{\mathfrak{F}}}[\{f(k)\}].$$

Now using the fact that $N\Delta x = T$ and $\Delta w = 1/T$ we have

$$\left\{F\left(\frac{j}{T}\right)\right\} = T\overline{\overline{\mathfrak{F}}}[\{f(k)\}].$$

Finally, application of Equation (5.12) yields

(5.14) $\{C_j\} = \overline{\overline{\mathfrak{F}}}[\{f(k)\}].$

Thus we see that the Fourier series coefficients are equal to the terms of the discrete Fourier transform of the sampled sequence *when the sampling rate is selected as per Equation* (5.13). Values of the coefficients for negative indices $(-j)$ are obtained by means of Theorem 4.3:

$$C_{-j} = C_{N-j}.$$

It is important to note that although there are an infinite number of Fourier series coefficients C_j, the method just presented will only yield the first N of them. However, inasmuch as all real world functions are almost band-limited, we can always choose a sufficiently large value of N to give a reasonable representation of the function.

SAMPLING BY MEANS OF THE COMB FUNCTION

In this section we present a brief but useful discussion of how sampling may be described in terms of the product of the original function and a comb function. Although this point of view leads to the same basic results that we have already demonstrated, it also permits additional insight into the sampling problem.

We define the comb function as an infinite train of equally spaced (Δx) impulse functions:

$$(5.15) \qquad \text{comb}_{\Delta x}(x) = \sum_{k=-\infty}^{\infty} \delta(x - k\Delta x).$$

This function is illustrated in Figure 5.15. A very important fact concerning the comb function is that its Fourier transform is also a comb function. More specifically we have the following.

Theorem 5.3. The comb function with spacing factor Δx,

$$\text{comb}_{\Delta x}(x) = \sum_{k=-\infty}^{\infty} \delta(x - k\Delta x),$$

has a Fourier transform that is also a comb function with spacing factor $\Delta w = 1/\Delta x$:

$$\mathcal{F}\left[\text{comb}_{\Delta x}(x)\right] = \text{comb}_{1/\Delta x}(w) = \sum_{j=-\infty}^{\infty} \delta\left(w - \frac{j}{\Delta x}\right).$$

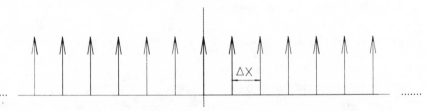

FIGURE 5.15. Comb function.

This theorem can only be rigorously proven using distribution theory which, unfortunately, is beyond the scope of this text.

We now demonstrate that the product of a function $h(x)$ and the shifted delta function $\delta(x - a)$ is equal to the shifted delta function scaled by a constant equal to the value of the function at a:

$$(5.16) \qquad h(x)\delta(x - a) = h(a)\delta(x - a).$$

Again, this result can only be rigorously established by means of distribution theory. However, we "illustrate" the proof using properties of the delta function. We begin by considering the integral of the product of $h(x)$ and $\delta(x - a)$:

$$\int_{-\infty}^{\infty} h(x)\delta(x - a)\, dx.$$

Recall from Chapter 3 that $\delta(x - a)$ is everywhere zero except at the point a. Thus the preceding integral becomes

$$\int_{a^-}^{a^+} h(x)\delta(x - a)\, dx.$$

However, over the infinitesimally small domain $[a^-, a^+]$ $h(x)$ is essentially a constant equal to $h(a)$. Thus we can remove it from inside the integral sign:

$$\int_{a^-}^{a^+} h(x)\delta(x - a)\, dx = h(a)\int_{a^-}^{a^+} \delta(x - a)\, dx = h(a)\int_{-\infty}^{\infty} \delta(x - a)\, dx.$$

Therefore, we see that the integral of the product of $h(x)$ and $\delta(x - a)$ is the same as the product of the constant $h(a)$ and the integral of $\delta(x - a)$. From this we infer our desired result.

In a similar way we can generalize this result to read

$$h(x)\sum_{k=-\infty}^{\infty} \delta(x - k\Delta x) = \sum_{k=-\infty}^{\infty} h(k\Delta x)\delta(x - k\Delta x),$$

or in terms of the comb function

$$(5.17) \qquad h(x)\mathrm{comb}_{\Delta x}(x) = \sum_{k=-\infty}^{\infty} h(k\Delta x)\delta(x - k\Delta x).$$

Let us consider the following interpretation of Equation (5.17). When we multiply a function $h(x)$ by $\mathrm{comb}_{\Delta x}(x)$ we, in fact, form a train of "weighted" impulse functions. The weight, or strength, of each $\delta(x - k\Delta x)$ term is determined by the value of the function $h(x)$ at that location, that is, $h(k\Delta x)$. This is, in effect, a mapping of the set of integers to the set of numbers $\{h(k\Delta x)\}$ or, in other words, a definition of the sequence $\{h(k\Delta x)\}$. However, this coincides with our concept of sampling the function $h(x)$ with sampling rate Δx. Thus $h(x)\text{comb}_{\Delta x}(x)$ may be considered a compact mathematical description of sampling. In fact $h(x)\text{comb}_{\Delta x}(x)$ may be considered either a sequence or a function (more properly a distribution).

Let us now assume that we have a band-limited function $f(x)$ (bandwidth 2Ω) that we desire to sample. We have just illustrated that the sampled sequence $\{h(k\Delta x)\}$ can be written as

$$h(x) = \{h(k\Delta x)\} = f(x)\text{comb}_{\Delta x}(x).$$

Taking the Fourier transform of this equation and using the product Theorem 3.10 we obtain

$$(5.18) \qquad H(w) = \mathscr{F}\left[f(x)\text{comb}_{\Delta x}(x)\right] = F(w) * \text{comb}_{1/\Delta x}(w).$$

However, since the impulse function is the unit element under convolution (see Equation 3.24), the preceding equation becomes

$$(5.19) \qquad H(w) = \sum_{j=-\infty}^{\infty} F\left(w - \frac{j}{\Delta x}\right).$$

Thus we see that the Fourier transform of the sampled sequence $\{h(k)\}$ is simply the Fourier transform of the envelope function $h(t)$ periodically displaced by an amount $1/\Delta x$ (see Figure 5.16). We can see from this figure that if the period $1/\Delta x$ is greater than the bandwidth 2Ω, then these transforms will not overlap or interfere, and we can uniquely obtain the transform $F(w)$ by

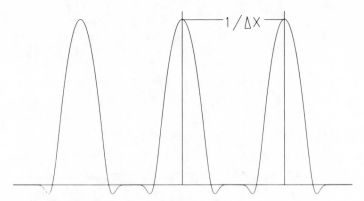

FIGURE 5.16.   Fourier transform of the sampled function.

considering $H(w)$ (the transform of the sampled sequence) from $-\Omega$ to Ω. Note that this condition for no overlap is simply

$$\frac{1}{\Delta x} > 2\Omega \quad \text{or} \quad \Delta x < \frac{1}{2\Omega}.$$

SUPER-GAUSSIAN WINDOWS

At the beginning of this chapter we noted that when sampling a function we had to concern ourselves with windowing. That is, we had to be careful to sample the function over its entire domain of interest. In Figure 5.17 we show an example of a function that has been windowed. As can be seen in this figure, the original function extends over the domain $[-X, X]$ and it has only been sampled over the subdomain $[-A, A]$. In this case the result is sharp discontinuities at the endpoints $-A$ and A. As we learned in Chapter 2, the Fourier series representation is very sensitive to such discontinuities which cause Gibbs effect or ringing. As it turns out, the Fourier transform will also tend to ring when sharp discontinuities are encountered (consider, for example, the transform of a pulse function).

In actual practice, memory size limitations of the digitizing (and recording) equipment often make it impossible to properly sample a function over its entire domain. Consequently, window effects or sharp discontinuities become unavoidable. When this is the case, the common "solution" is to place an

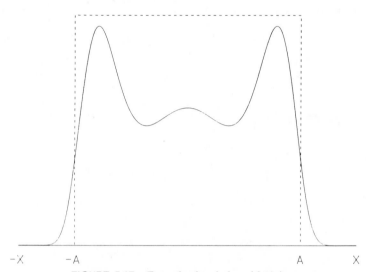

FIGURE 5.17.   Example of a windowed function.

artificial window over the function that *gradually* reduces the function to zero at the endpoints. In figurative terms we use an artificial window with rounded corners. There are many such windows such as the Kaiser, Lanczos, hamming, and hanning to name just a few. Unfortunately, there is no clear cut or overall theory that describes windowing but instead its use is as much an art as it is a science.

In this section we discuss super-Gaussian windows that were first physically realized as apodized apertures (at Lawrence Livermore National Laboratory) and used to minimize diffraction effects in high-powered laser beams used in fusion research. Before we present their formal definition, let us consider the problem of windowing in mathematical terms. We begin with a time-limited function $f(t)$ over the domain $[-X, X]$ whose Fourier transform is denoted as $F(w)$. When we window a function such as the one shown in Figure 5.17, we multiply it by a pulse function of half-width A:

$$g(x) = f(x)p_A(x).$$

Therefore, when we transform this truncated function $g(x)$ instead of $F(w)$ (our desired result), we obtain $F(w)$ convolved with, or smeared by, $2A \operatorname{sinc} 2\pi wA$ which is the transform of $p_A(x)$ (see Theorem 3.10). That is to say, we obtain

(5.20)

$$G(w) = 2AF(w) * \operatorname{sinc}(2\pi wA) = 2A \int_{-\infty}^{\infty} F(\xi)\operatorname{sinc}[2\pi A(w - \xi)]\, d\xi.$$

In Figure 5.18 we show both $F(\xi)$ and $\operatorname{sinc}[2\pi A(w - \xi)]$. As we learned in Chapter 3, the value of the convolution function $G(w)$ (for a particular value

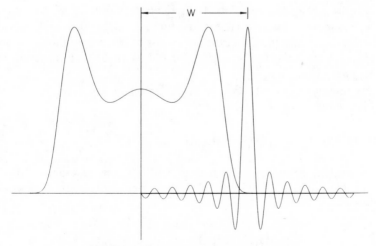

FIGURE 5.18. $F(w)$ and $\operatorname{sinc}[2\pi A(w - \xi)]$.

of w) is equal to the area under the product curve of the two functions being convolved. As can be appreciated from this figure, the resulting product is dependent upon the separation w and the relative widths of $\operatorname{sinc} 2\pi wA$ and $F(w)$. Because of all these factors we can now begin to appreciate why there is no specific theory of windowing. We can only look for trends. For example, if the window is wide (i.e., A is large) relative to the time limits of $f(x)$, then the width of $\operatorname{sinc} 2\pi wA$ will be small relative to the width of $F(w)$. Consequently, windowing problems will be less. As a matter of fact, in the limiting case as A approaches infinity, $\operatorname{sinc} 2\pi wA$ becomes a delta function of strength $1/2A$. In this case Equation (5.20) becomes

$$G(w) = \frac{2AF(w) * \delta(w)}{2A} = F(w) * \delta(w) = F(w),$$

and as expected there are no windowing problems at all.

To help illustrate these concepts we choose a specific function, namely, the Gaussian

$$f(x) = e^{-x^2},$$

and place three different windows over it. The first window (see Figure 5.19a) is such that it clips, or apertures, the function at the 0.25 level. That is, the half-width A of the window is chosen such that

$$f(A) = e^{-A^2} = 0.25 \quad \text{or} \quad A = 1.18.$$

Shown in Figure 5.19b is the true transform $F(w)$ and the $\operatorname{sinc} 2\pi wA$ function. Finally in Figure 5.19c we show the calculated transform $G(w)$, that is, the convolution of $F(w)$ and $\operatorname{sinc} 2\pi wA$. In Figures 5.20 and 5.21 we show this same sequence for windows that clip the function at the 0.1 and 0.005 levels, respectively. In these figures the relative widths of $F(w)$ and $\operatorname{sinc} 2\pi wA$ should be noted with attention to the amount of Gibbs effect that is present in the calculated transform $G(w)$.

Let us now turn our attention to super-Gaussian windows. We define the super-Gaussian, or simply SG, function as

(5.21) $$SG(x; A, N) = e^{-|x/A|^N},$$

where N is called the power and A the half-width. Let us first take a closer look at the meaning of the half-width A. We note that, for all powers of N, when $x = A$ we have

$$SG(A; A, N) = e^{-|A/A|^N} = \frac{1}{e}.$$

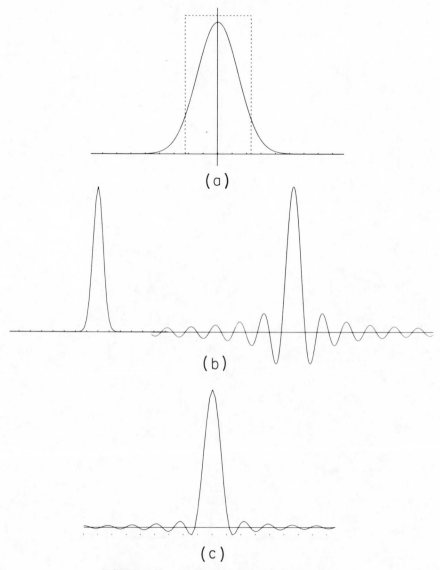

FIGURE 5.19. Function windowed at the 0.25 level.

Therefore, we see that A is, in fact, the half-width of the SG function out to the $1/e = 0.3678 \cdots$ location. In Figure 5.22 we show the SG functions for several different powers of N. When $N = 2$ the SG function becomes a simple Gaussian function:

$$\mathrm{SG}(x; A, 2) = e^{-(x/A)^2}.$$

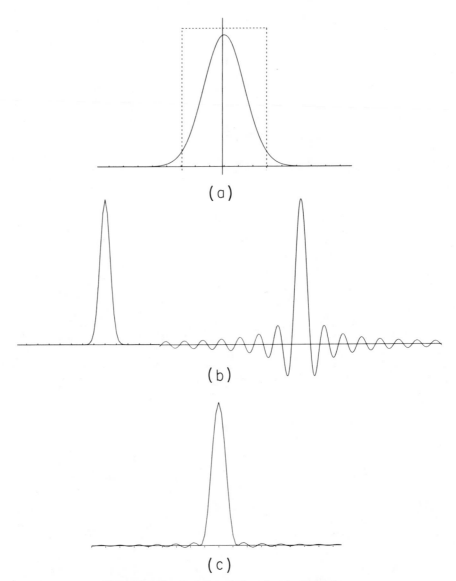

FIGURE 5.20.   Function windowed at the 0.1 level.

In the limit as N approaches infinity the SG function becomes a pulse function of half-width A:

$$\lim_{N \to \infty} \text{SG}(x; A, N) = p_A(x).$$

For all values of N between 2 and ∞ the SG function may be considered a pulse function with rounded corners (at both the top and bottom). Thus we can see why SG functions make ideal artificial windows. In other words, when

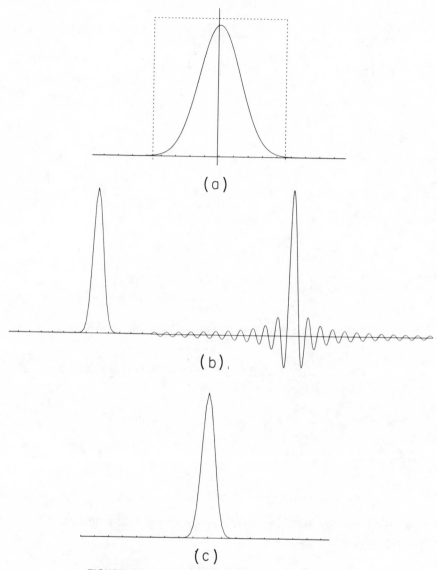

FIGURE 5.21. Function windowed at the 0.005 level.

we multiply a function $f(x)$ by a SG function (with the proper values of A and N), we can gradually reduce the function to zero and alleviate the effects of sharp discontinuities.

Actually, the SG functions never exactly equal zero but instead approach zero in the limit:

$$\lim_{x \to \infty} SG(x; A, N) = 0.$$

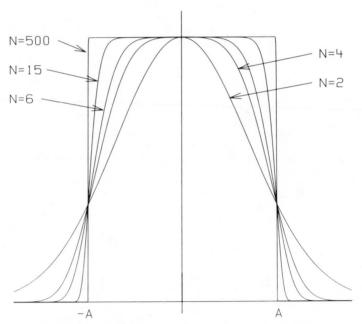

FIGURE 5.22.   Super-Gaussian functions for various values of N.

Mathematically, we write: for all finite values of $N > 0$ and $A > 0$, given any $\varepsilon > 0$ there exists a value $C > A$ such that

$$|\text{SG}(x; A, N)| = e^{-|x/A|^N} < \varepsilon \quad \text{for all } |x| \geqslant C.$$

Solving this equation for C we obtain

(5.22) $$C \geqslant A(-\ln \varepsilon)^{1/N}.$$

Now using the SG function in place of the pulse function as a window, Equation (5.20) becomes

$$G(w) = F(w) * \mathcal{F}[\text{SG}(x; A, N)].$$

In other words, the calculated Fourier transform $G(w)$ is now given as the convolution of the true transform $F(w)$ and the transform of the SG window $\mathcal{F}[\text{SG}(x; A, N)]$. For the two limiting cases of $N = 2$ and $N = \infty$ we are able to write analytical expressions for the transforms:

(5.23) $$\mathcal{F}[\text{SG}(x; A, 2)] = (\pi)^{1/2} A e^{-(\pi A w)^2},$$

(5.24) $$\mathcal{F}[\text{SG}(x; A, \infty)] = 2A \operatorname{sinc} 2\pi wA.$$

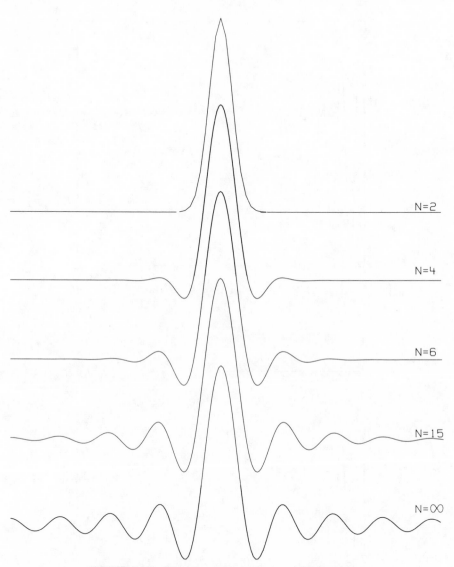

FIGURE 5.23. Fourier transforms of selected SG functions.

The transforms for values of N between 2 and ∞ must be digitally obtained. In Figure 5.23 we show several such transforms (actually the transforms of the functions shown in Figure 5.22). Note that as the value of N increases, so too does the amount of the ringing. As can be seen in this figure, the low frequency portions of the transforms are approximately equal. To be more specific, we characterize the width by the distance to the $1/e = 0.3678$ location. It is a trivial matter to solve Equation (5.23) for the case in which $N = 2$ to determine

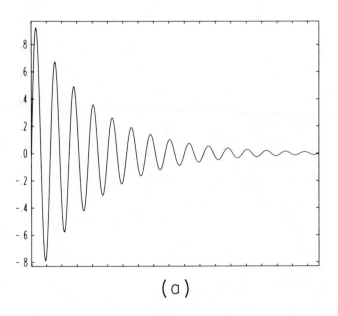

(a)

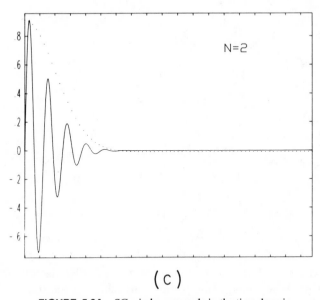

N=2

(c)

FIGURE 5.24. SG window example in the time domain.

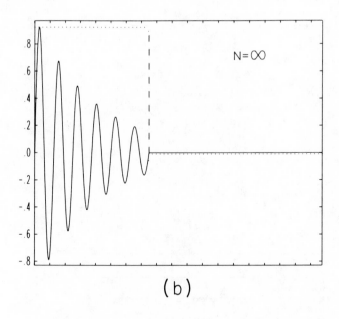

(b)

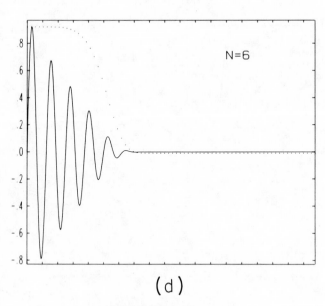

(d)

FIGURE 5.24. (*Continued*).

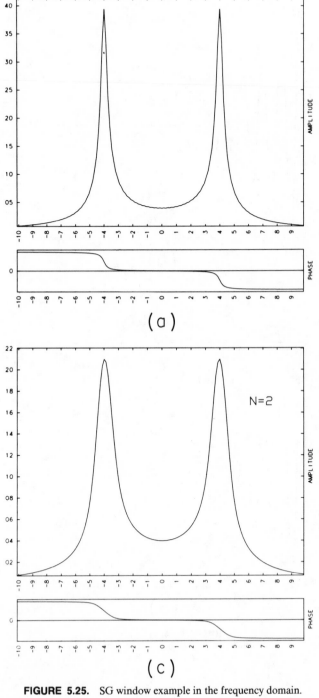

FIGURE 5.25. SG window example in the frequency domain.

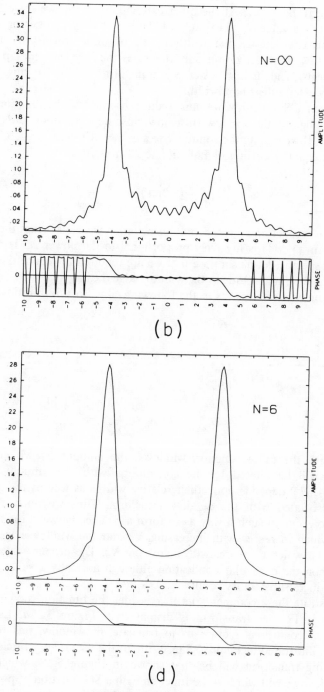

(b)

(d)

FIGURE 5.25. (*Continued*).

this distance to be $1/\pi A$. At the other extreme ($N = \infty$), we solve Equation (5.24) to obtain a value equal to $1.09/\pi A$. Thus with only slight error (less than 10% we can characterize the width of all of these transforms as $1/\pi A$. Obviously, this characterization only describes the low frequency portion of the transforms. The higher power SG transforms exhibit ringing that can significantly extend their bandwidth.

When using a SG window we must reduce the value of the function close to zero at the edge of the window (just how close depends upon the particular application). If we denote the window edges as $x = +C$ and $x = -C$, then we can use the equality portion of Equation (5.22) to write

$$\frac{C}{A} = (-\ln \varepsilon)^{1/N},$$

where ε is the reduction factor. From this equation we can see that the width of the transform ($1/\pi A$) is inversely proportional to the power N. To be somewhat more specific, let us suppose that we are required to reduce the function by three orders of magnitude at the window edge (i.e., $\varepsilon = 0.001$). For this case the preceding equation results in the following edge–to–width ratios:

N	$\dfrac{C}{A}$	N	$\dfrac{C}{A}$
2	2.63	6	1.38
3	1.90	7	1.32
4	1.62	15	1.14
5	1.47	∞	1.00

Just as with the pulse function windows, the amount by which the true transform will be altered by the convolution with the window transform $\mathcal{F}[SG(x; A, N)]$ depends upon their relative widths as well as the amount of ringing associated with the window transform. Thus we must choose the window power N in such a way as to form a balance between the amount of ringing (which increases with increasing N) and the window edge to width ratio (C/A) (which decreases with increasing N). The proper value of N will depend upon the particular application although normally a value between 4 and 8 produces satisfactory results.

To illustrate these comments let us consider the function shown in Figure 5.24a whose Fourier transform is displayed in Figure 5.25a. Let us now assume that conditions require us to truncate, or window, this function at $x = 1.5$ as illustrated in Figure 5.24b. Without the use of an artificial window, the resulting transform will be that shown in Figure 5.25b which contains ringing. Next we will window the function with a SG function of power $N = 2$ and reduction factor of $\varepsilon = 0.001$. In this case, the edge–width ratio is 2.63 and thus the window width A is chosen to be equal to 0.57. This window is shown in Figure 5.24c and the resulting transform in Figure 5.25c. Note that in this case the ringing has been eliminated; however, the width of the transform peaks have been significantly increased. Finally, as a compromise we choose a power of N equal to 6 which requires a window width of $A = 1.09$. This is illustrated in Figure 5.24d and the resulting transform in Figure 5.25d. Note that the ringing has been significantly reduced as has the widening of the peaks.

SUMMARY

In this chapter we have presented methods that permit the computer calculation of the Fourier series and Fourier transform of a function. In each situation the procedure is similar. First the function in question is sampled and converted to a sequence. Then the discrete Fourier transform of this sequence is calculated. The Fourier transform can be recovered by passing a smooth curve through the points of the discrete Fourier transform sequence. The Fourier series coefficients, on the other hand, are given as the terms of the discrete Fourier transform sequence when the sampling rate is chosen properly. We also presented a brief discussion of the theory of sampling and windowing.

BIBLIOGRAPHY

Freeman, H., *Discrete Time Systems, An Introduction to the Theory*, Wiley, New York, 1965.

Raven, F. H., *Automatic Control Engineering*, McGraw-Hill, New York, 1968.

Rabiner, L., and B. Gold, *Theory and Applications of Digital Signal Processing*, Prentice-Hall, New York, 1975.

Oppenheim, A. V. and R. Schafer, *Digital Signal Processing*, Prentice-Hall, New York, 1975.

Arsac, J., *Fourier Transforms and the Theory of Distributions*. Prentice-Hall, Englewood Cliffs, N.J., 1966.

Papoulis, A., *Systems and Transforms with Applications in Optics*, McGraw-Hill, New York, 1968.

CHAPTER

6

SYSTEMS AND
TRANSFER FUNCTIONS

To truly appreciate the application of Fourier analysis to the various fields of science and engineering we must become somewhat familiar with the basic theory of transfer functions. Transfer function theory plays an important and powerful role in the field of systems theory. It is the link that connects Fourier analysis (as well as Laplace transform analysis) with the impulse response of a system. It may also be considered a physical application of the convolution Theorem 3.9. In this chapter we present a brief discussion of a system as well as the transfer function and impulse response of a system.

CONCEPT OF A SYSTEM

In this section we discuss the concept of linear and invariant systems. In general, *a system is defined to be a mapping of a set of input functions to a set of output functions.* An equivalent but less compact description is: *a system is defined as a mathematical abstraction devised to serve as a model for physical phenomenon. It consists of an input function $f(t)$, an output function $y(t)$, and a cause–effect relationship between them.*

149

The following notation is often used to denote this cause–effect relationship:

(6.1) $$y(t) = S[f(t)].$$

Another commonly used notation is

(6.2) $$S: f(t) \rightarrow y(t).$$

As an example, let us consider a system that produces an output function that is simply the square of the input function. This system would be denoted as:

$$S: f(t) \rightarrow f^2(t),$$

or

$$y(t) = f^2(t).$$

Often a system will make use of operations from calculus such as the derivative and integral when mapping the input function to the output function. For example, consider the system defined as

$$S: f(t) \rightarrow f'(t) + f''(t) + \int_0^t f(\xi)\, d\xi.$$

If we were to use the function $f(t) = t^2$ as input to this system, then the resulting output would be $y(t) = t^3/3 + 2t + 2$.

Let us now limit our attention to single-input single-output systems that are linear and invariant. We say a system is *linear* if and only if

$$S(af_1(t) + bf_2(t)) = aS(f_1(t)) + bS(f_2(t)),$$

where a and b are arbitrary (perhaps complex) constants. A system is said to be *invariant* (or *stationary*) if and only if for any value τ

$$S(f(t - \tau)) = y(t - \tau).$$

Physically speaking, a system is linear if a linear combination of inputs yields a linear combination of the respective outputs. That is, let the response of the system to input $f_1(t)$ be $y_1(t)$ and to $f_2(t)$, let it be $y_2(t)$. Next as input to the system, let us use $af_1(t) + bf_2(t)$. Then the system is linear if and only if the resulting output is $ay_1(t) + by_2(t)$. For example, the system

$$S: f(t) \rightarrow f'(t) + 2f(t)$$

is linear because

$$S(af_1(t) + bf_2(t)) = af_1'(t) + 2af_1(t) + bf_2'(t) + 2bf_2(t),$$

$$= aS[f_1(t)] + bS[f_2(t)].$$

Physically speaking, a system is invariant if a displaced or shifted input function $f(t - \tau)$ will yield the output function with the same displacement or shift $y(t - \tau)$.

In our work we deal exclusively with systems that can be described by a linear combination of the calculus operators. In other words, we only deal with systems that can be described by linear differential equations. To facilitate our work we introduce the Heaviside operator, denoted as s^i, and defined as

$$s^i f(t) = \frac{d^i f(t)}{dt^i} = f^{[i]}(t),$$

$s^{-i}f(t)$ is the ith integral of $f(t)$. For example,

$$s^2 f(t) = \frac{d^2 f(t)}{dt^2} = f^{[2]}(t),$$

$$s^{-2}f(t) \int_0^t \left[\int_0^\eta f(\xi) \, d\xi \right] d\eta.$$

The special case of $i = 0$ is defined as $s^0 f(t) = f(t)$.

As it turns out, the Heaviside operator may be treated as an algebraic quantity and added, subtracted, and multiplied, and sometimes divided.[†]

With the definition of the Heaviside operator in hand and our previously mentioned assumption that the system can be described by a linear differential equation in terms of the output $f(t)$ we are able to rewrite Equation (6.1) as

(6.3)
$$y(t) = \sum_{i=0}^{k} a_i s^i f(t),$$

or

(6.4)
$$S : f(t) \rightarrow \sum_{i=0}^{K} a_i s^i f(t),$$

where the a_i terms are constants. For example, to describe the system

$$y(t) = f'(t) + 2f(t)$$

[†] For more information on this see texts by Raven listed in the Bibliography at the end of this chapter.

in terms of Equation (6.3) we would write

$$y(t) = sf(t) + 2f(t) = (s + 2)f(t).$$

Note that Equation (6.3) describes a system in which the output $y(t)$ is only dependent upon the input $f(t)$ [and derivatives of $f(t)$]. We now generalize our system definition to include systems in which a portion of the output and its derivatives are used as feedback to help determine the new output. These (feedback) systems are described by differential equations involving both the input $f(t)$ and the output $y(t)$:

$$(6.5) \qquad \sum_{i=0}^{M} b_i s^i y(t) = \sum_{i=0}^{K} a_i s^i f(t).$$

A feedback system is linear and invariant in terms of the quantity $(b_0 + b_1 + \cdots + b_M)y(t)$ rather than just $y(t)$.

IMPULSE RESPONSE OF A SYSTEM

In this section we demonstrate the fact that a linear invariant system can be completely characterized by a knowledge of how it responds when the input function is an impulse, or delta, function.

Let S be a system that maps a general input function $f(x)$ to the corresponding output function $y(x)$. The *impulse response* of such a system, denoted as $h(x)$, is defined as the output that results when the system is excited with a delta or impulse function:

$$h(x) = S[\delta(x)]$$

By assumption, S is both linear and invariant and, consequently, an input of the form $a\delta(x) + b\delta(x - \tau)$ will result in the output function $ah(x) + bh(x - \tau)$. Shown in Figure 6.1 is an arbitrary function $f(x)$ whose domain has been divided into equal intervals of width Δx. As indicated in this figure, at every location $k\Delta x$ a rectangle of height $f(k\Delta x)$ and width Δx is constructed. Now as input to our system S, let us consider a linear combination of shifted impulses $\delta(x - k\Delta x)$, each of which is weighted by the constant $f(k\Delta x)\Delta x$:

$$(6.6) \qquad g(x) = \sum_{k=-\infty}^{\infty} f(k\Delta x)\Delta x\delta(x - k\Delta x).$$

Again linearity and invariance of the system guarantee that the output must be the same linear combination of the shifted impulse response functions:

$$(6.7) \qquad y(x) = \sum_{k=-\infty}^{\infty} f(k\Delta x)\Delta x h(x - k\Delta x).$$

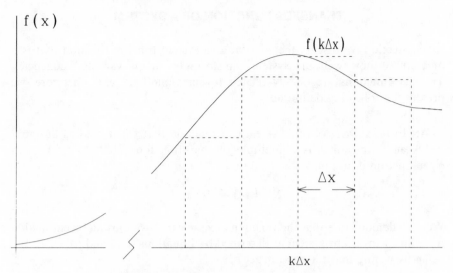

FIGURE 6.1. Arbitrary function for convolution.

Essentially, what we have is an infinite train of weighted impulses exciting the system, with the resulting output being a linear combination of the responses to each of these impulses. If now in Equations (6.6) and (6.7) we "shrink" the interval width Δx to zero and use a somewhat heuristic calculus approach, we can reason that $k\Delta x$ approaches a continuous variable ξ, Δx becomes $d\xi$, and the summation goes to an integral. Equation (6.6) then becomes

$$g(x) = \int_{-\infty}^{\infty} f(\xi)\delta(x - \xi)\, d\xi = f(x) * \delta(x) = f(x).$$

We see, therefore, that (in the limit) our input train of weighted impulses equals the general function $f(x)$. Using similar reasoning we can demonstrate that Equation (6.7) becomes

(6.8) $$y(x) = \int_{-\infty}^{\infty} f(\xi)h(x - \xi)\, d\xi = f(x) * h(x).$$

Equation (6.8) tells us that the output of a system $y(x)$, resulting from any general input function $f(x)$, is simply the convolution of that input function with the impulse response function of the system. Therefore, if we know how a system responds when excited by an impulse, then we can use Equation (6.8) to calculate its response to any general input function.

TRANSFER FUNCTION OF A SYSTEM

In this section we discuss the concept of a system transfer function that is a measure of how well the system transforms or maps various frequencies. Transfer functions are closely related to eigenfunctions, and, therefore, we present the following definition:

AN EIGENFUNCTION of a system is a function that when used as an input yields an output that is a multiple of that function. That is, $e(x)$ is an eigenfunction if and only if

$$S(e(x)) = Ae(x).$$

We now demonstrate that the function $e(x) = \exp(2\pi iwx)$ is an eigenfunction for a linear invariant system as described by Equations (6.3) and/or (6.4). We begin by noting that for any integer j

$$s^j \exp(2\pi iwx) = s^j e^{2\pi iwx} = (2\pi iw)^j e^{2\pi iwx},$$

where s^j is the previously defined Heaviside operator. If now in Equation (6.4) we let

$$f(x) = e(x) = e^{2\pi iwx},$$

we obtain

$$S: e(x) \rightarrow \sum_{j=0}^{K} a_j (2\pi iw)^j e(x).$$

Therefore, we see that $e(x) = \exp(2\pi iwx)$ is an eigenfunction for linear invariant continuous systems. That is, the output function is simply equal to the input function multiplied by

(6.9) $$H(w) = \sum_{j=0}^{K} a_j (2\pi iw)^j.$$

We define $H(w)$ to be the *transfer function* of a linear invariant continuous system. Note that $H(w)$ is a constant with respect to the variable x, although we soon consider it to be a function of the variable w. As an example, consider the system defined as

$$S: f(x) \rightarrow (s^2 + 3s + 2)f(x);$$

then

$$H(w) = (2\pi i w)^2 + 3(2\pi i w) + 2.$$

For any particular value $w = w_1$, recall (see Chapter 1) that is $\exp(2\pi i w_1 x)$ is a single frequency sinusoid. Inasmuch as it is an eigenfunction, when it is used as input to a linear invariant system, the output will also be a sinusoid of the same frequency. However, the amplitude of the output will differ from the input by the factor $H(w_1)$. Thus we see that the transfer function is a measure of how well a system passes frequency terms. To clarify these comments let us be more specific for a moment and consider a system whose transfer function is described as

$$H(w) = 2w.$$

If we excite this system with a single frequency term of $w = 1$ cycle [$\exp(2\pi i x)$] then the output frequency term will also be 1 cycle. However, it will have an amplitude equal to twice that of the input. Similarly, a single frequency input of $w = 2$ cycles [$\exp(4\pi i x)$] will result in an output term of 2 cycles with an amplitude of four times that of the input.

By assumption, the systems being considered are linear and invariant and, consequently, if the input consists of two single frequency terms then the output will also consist of two single frequency terms, each modified by the transfer function evaluated at those specific frequencies:

$$S : (e^{2\pi i w_1 x} + e^{2\pi i w_2 x}) \rightarrow H(w_1)e^{2\pi i w_1 x} + H(w_2)e^{2\pi i w_2 x}.$$

As we have seen in previous chapters, any function can be written as a linear combination of these single frequency terms. In particular, a periodic function can be written as a Fourier series:

$$f(x) = \sum_{k=-\infty}^{\infty} C_k e^{2\pi i w_k x}.$$

If this function is now used as input to our system, then the output is given as

$$y(x) = \sum_{k=-\infty}^{\infty} H(w_k) C_k e^{2\pi i w_k x}.$$

This equation is actually a Fourier series representation of the output function $y(x)$. Thus we see that the transfer function modifies the Fourier series coefficients of the input function C_k to obtain the Fourier series coefficients of the output function $H(w_k)C_k$. In other words, the transfer function converts the frequency content of the input function to that of the output function.

Let us now consider the effect of the transfer function upon the Fourier transform of a function. A (well-behaved) arbitrary function may also be represented by its Fourier transform frequency content (see Chapter 3):

$$f(x) = \int_{-\infty}^{\infty} F(w) e^{2\pi i w x} \, dw.$$

If this function is used to force the system, then each individual frequency term $\exp(2\pi i w x)$ will be modified by the transfer function $H(w)$ evaluated at that frequency. Since the system is linear and invariant, the output will be the summation (integral) of all these output terms:

(6.10)
$$y(x) = \int_{-\infty}^{\infty} H(w) F(w) e^{2\pi i w x} \, dw.$$

A close look at the preceding equation reveals that it is, in fact, the Fourier transform representation of the output function $y(x)$. Thus we see that the transfer function modifies the Fourier transform (or frequency content) of the input function $F(w)$ to yield that of the output function $H(w)F(w)$.

IMPULSE RESPONSE AND TRANSFER FUNCTIONS

In this section we demonstrate that the transfer function of a system is simply the Fourier transform of its impulse response. We begin by using the convolution theorem 3.9 and Fourier transforming both sides of Equation (6.8) to obtain

$$Y(w) = F(w) \mathcal{H}(w),$$

where $Y(w)$ is the Fourier transform of $y(x)$, $F(w)$ is the Fourier transform of $f(x)$, and $\mathcal{H}(w)$ is the Fourier transform of $h(x)$. If we now take the inverse Fourier transform of this equation, we obtain

$$y(x) = \int_{-\infty}^{\infty} \mathcal{H}(w) F(w) e^{2\pi i w x} \, dw.$$

Comparing this equation to Equation (6.10) we easily see that the transfer function $H(w)$ is, in fact, the Fourier transform of the impulse response function of the system. A little thought reveals that this result should not be too surprising. The transfer function of a system is a measure of how the system responds to any, and all, frequencies. To determine $H(w)$ we must excite the system with all possible frequencies. In Chapter 3 we saw that the impulse function contains all frequencies:

$$\mathscr{F}[\delta(x)] = T(w) = 1.$$

Thus when we excite a system with an impulse (in the "time" domain) we are, in fact, exciting it with all frequencies (in the frequency domain).

SUMMARY

In this chapter we presented some basic theory for single-input single-output linear invariant systems. In doing so we presented both a "time" domain (impulse response) and a frequency domain (transfer functions) analysis. We learned that the transfer function of a system was simply the Fourier transform of the impulse response function. It must be pointed out that in this chapter we presented only a certain limited view of the systems theory and this presentation should in no way be considered a thorough treatment of the subject. It does, however, serve as a foundation upon which to construct the remaining chapters in this text, in which we introduce additional systems concepts as they are required.

BIBLIOGRAPHY

Freeman, H., *Discrete Time Systems, An Introduction to the Theory*, Wiley, New York, 1965.

Raven, F. H., *Mathematics of Engineering Systems*, McGraw-Hill, New York, 1966.

Goodman, J. W., *Introduction to Fourier Optics*, McGraw-Hill, New York, 1968.

Athans, M., and P. L. Falb, *Optimal Control*, McGraw-Hill, 1966.

Raven, F. H., *Automatic Control Engineering*, McGraw-Hill, 1968.

CHAPTER

$$7$$

VIBRATIONAL SYSTEMS

In this chapter we take a look at mechanical vibrational systems, which consist of springs, masses, and dampers, as well as electrical vibrational systems which consist of resistors, capacitors, and inductors. Although these two systems are physically quite different, they are both described mathematically by the same type of linear differential equation. We show how to determine these differential equations and from them construct the transfer functions of the systems. Later in the chapter, we look at the wave equation and study the vibrational transverse motion of an elastic string.

TRANSFER FUNCTION OF THE GENERAL EQUATION

The general differential equation that describes the behavior of a mechanical or electrical vibrational system is

$$(7.1) \qquad \sum_{k=0}^{K} b_k s^k y(t) = \sum_{j=0}^{J} a_j s^j f(t),$$

where s^k is the Heaviside operator defined in Chapter 6; $f(t)$ is considered the driving function and $y(t)$ the resulting output. For mechanical systems, $f(t)$ usually takes on the form of a driving force or driving motion and $y(t)$ an output displacement, velocity, or acceleration. For electrical systems, $f(t)$ is usually a driving voltage or current and the output $y(t)$ is some resulting voltage or current.

We first determine the transfer function of Equation (7.1). What we wish to obtain is $H(w)$ such that, when $f(t)$ equals the eigenfunction $\exp(2\pi iwt)$, then $y(t) = H(w)\exp(2\pi iwt)$. We begin by substituting

$$f(t) = e(t) = e^{2\pi iwt}$$

into Equation (7.1). This results in

(7.2)
$$\sum_{k=0}^{K} b_k s^k y(t) = A(w)e(t),$$

where

(7.3)
$$A(w) = \sum_{j=0}^{J} a_j (2\pi iw)^j.$$

Now if we let $y(t) = H(w)e(t)$ and substitute this into Equation (7.2) we obtain

(7.4)
$$G(w)H(w)e(t) = A(w)e(t),$$

where

(7.5)
$$G(w) = \sum_{k=0}^{K} b_k (2\pi iw)^k.$$

Solving for the overall transfer function in Equation (7.4) we obtain

(7.6)
$$H(w) = \frac{A(w)}{G(w)}.$$

Thus we see that the overall transfer function $H(w)$ is simply the transfer function of the feedforward portion $A(w)$ divided by the transfer function of the feedback portion $G(w)$.

MECHANICAL SYSTEMS

In this section we determine the equations of motion of a spring, a mass, and a damper, as well as several (series and parallel) combinations of these elements.

The relationship between a force acting on a spring (Figure 7.1*a*) and the resulting displacement is given by

$$(7.7) \qquad\qquad f(t) = Kx(t),$$

where K is the spring constant. We may consider a spring to be a system in which the input is the force $f(t)$ and the output is the resulting displacement $x(t)$. Equation (7.7) is obviously a special case of Equation (7.1) in which

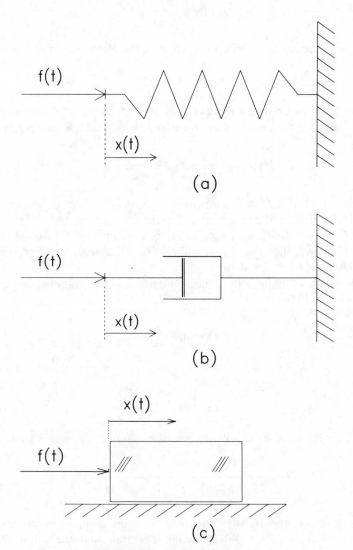

FIGURE 7.1. Mechanical elements.

$a_0 = 1$, $b_0 = K$, and all other constants are equal to zero. Direct application of Equations (7.3), (7.5), and (7.6) yields the transfer function

$$(7.8) \qquad\qquad\qquad H(w) = \frac{1}{K}.$$

The force acting on a damper (Figure 7.1b) is proportional to its velocity. Mathematically, we have

$$f(t) = C\frac{dx(t)}{dt},$$

where C is the damping constant. In terms of the Heaviside operator we have

$$(7.9) \qquad\qquad\qquad f(t) = Csx(t).$$

Just as with the spring, a damper may be considered a system with input $f(t)$ and output $x(t)$. The resulting transfer function of this damper system is

$$(7.10) \qquad\qquad H(w) = \frac{1}{2\pi iwC} = \frac{1}{2\pi wC}e^{-i\pi/2}.$$

We note that this transfer function is complex. That is to say, it has amplitude $1/2\pi wC$ and phase $-\pi/2$. From this equation and the remarks in Chapter 6 we see that when the input to a damper is sinusoidal, that is, $f(t) = \exp(2\pi iwt)$, then the output will also be sinusoidal; however, it will lag behind the input by $\pi/2$ radians.

The equation of motion of a mass (Figure 7.1c) is obtained by applying Newton's second law:

$$f(t) = M\frac{d^2x(t)}{dt^2}$$

or

$$(7.11) \qquad\qquad\qquad f(t) = Ms^2x(t).$$

From this equation we readily determine the transfer function to be

$$(7.12) \qquad\qquad H(w) = \frac{1}{(2\pi iw)^2 M} = \frac{1}{(2\pi w)^2 M}e^{-i\pi}.$$

Thus for a mass, the transfer function is again complex with amplitude $1/(2\pi w)^2 M$ and phase $-\pi$. When the input is sinusoidal, the output will also be sinusoidal and lag behind the input by π radians.

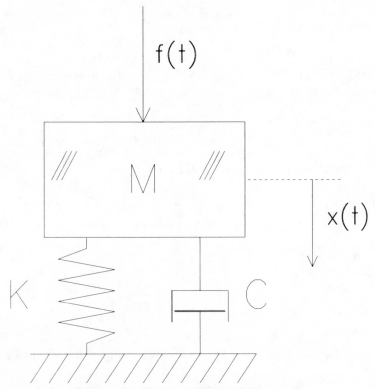

FIGURE 7.2. Mechanical damped oscillator system.

We now consider systems that are combinations of these three elements. As our first combination, let us study the classical system shown in Figure 7.2. Although there are several clever techniques available[†] for determining the equations of motion of this type of system, we use the less imaginative but more direct method of "balancing the forces." Note that when we perform this force balance we measure our displacement $x(t)$ from the equilibrium position. That is, we assume that the forcing function $f(t)$ is initially zero, and thus the weight of the mass will cause the spring and damper to deflect to some position $x(0)$. This is defined as the equilibrium position. As is obvious from the figure, the forces acting on the mass are the forcing function $f(t)$, the force due to the spring $-Kx(t)$, and the force due to the damper $-Csx(t)$. Newton's second law requires that when these forces are combined they must be equal to the resulting acceleration of the mass:

$$f(t) - Kx(t) - Csx(t) = Ms^2x(t),$$

[†]For example, see the text by Raven listed in the Bibliography at the end of this chapter.

or

(7.13)
$$\left(s^2 + \frac{C}{M}s + \frac{K}{M}\right)x(t) = \frac{f(t)}{M}.$$

Comparing this equation to Equation (7.1) we can easily write the transfer function of the system as

$$H(w) = \frac{1/M}{(2\pi iw)^2 + 2\pi iwC/M + K/M}$$

or, after a little complex variable algebra,

(7.14)
$$H(w) = \frac{1/M}{\sqrt{\left[K/M - (2\pi w)^2\right]^2 + (2\pi wC/M)^2}}e^{-\phi(w)}$$

where the phase $\phi(w)$ is given as

$$\tan[\phi(w)] = \frac{-2\pi wC/M}{K/M - (2\pi w)^2}.$$

Let us first consider the case of no damping (i.e., $C = 0$). When $C = 0$ is substituted into Equation (7.14) we obtain the undamped transfer function

(7.15)
$$H(w) = \frac{1/M}{K/M - (2\pi w)^2}.$$

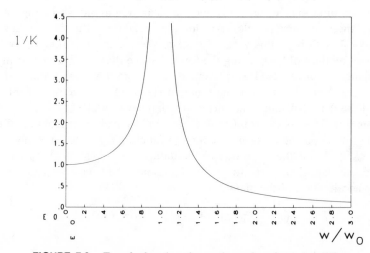

FIGURE 7.3. Transfer function of an undamped mechanical oscillator.

The natural frequency of this system is defined as

$$w_0 = \frac{(K/M)^{1/2}}{2\pi}$$

and, as is obvious from Equation (7.15), when w approaches this natural frequency the transfer function grows without bound. The magnitude of this transfer function is plotted versus w/w_0 in Figure 7.3. As the damping increases, the behavior of the transfer function in the neighborhood of the natural frequency improves. This is illustrated in Figure 7.4 for several values of C between 0 and $(KM)^{1/2}$. To gain additional insight into this system, let us be more specific and choose values for K, C, and M (i.e., $M = 1$ kg, $K = 4\pi^2$ N/m, and $C = 0.4(2\pi)$ N/m^2). When these values are substituted into Equation (7.14) we obtain

$$(7.16) \qquad H(w) = \frac{1}{4\pi^2\sqrt{w^4 - 1.84w^2 + 1}} e^{-i\phi(w)},$$

where

$$\tan[\phi(w)] = \frac{0.4w}{1 - w^2}.$$

The amplitude of this transfer function is shown in Figure 7.5a. We learned in Chapter 6 that the transfer function of a system is simply the Fourier transform of the impulse response function. Therefore, if we take the inverse

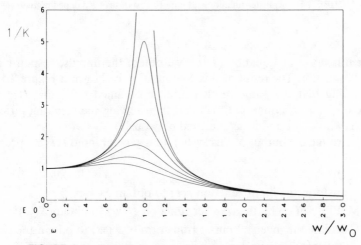

FIGURE 7.4. Transfer function of a damped mechanical oscillator.

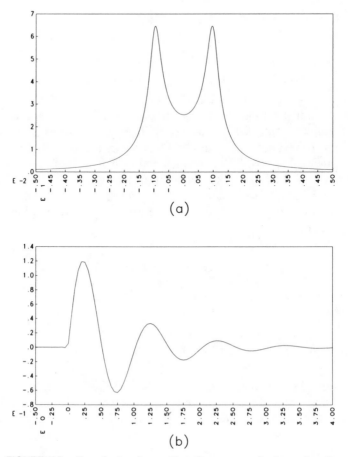

FIGURE 7.5. Transfer function and impulse response of a damped oscillator.

Fourier transform of Equation (7.16), we obtain the impulse response of this damped oscillator. The result of this computation is shown in Figure 7.5b. It is worth noting that this same result could be obtained by letting $f(t)$ be the impulse function in Equation (7.13) and then solving the resulting differential equation (with C, K, and M as specified previously).

To determine the output of this system to a general input $f(t)$ we proceed as follows:

1. Transform the input function $f(t)$ to obtain $F(w)$.
2. Multiply $F(w)$ by the transfer function $H(w)$ to obtain $X(w)$.
3. Calculate the inverse Fourier transform of $X(w)$ to obtain the resulting output $x(t)$.

For most "real world" problems, these transform operations must be performed on a digital computer using the techniques presented in Chapter 5. As an example, let us now determine the response of our system to the input function shown in Figure 7.6a. The Fourier transform of this function is shown in Figure 7.6b along with the transfer function (dashed line). The resulting product $X(w)$ is displayed in Figure 7.7a and, finally, the inverse transform of $X(w)$ [i.e., the output $x(t)$] is shown in Figure 7.7b.

Next let us consider the system shown in Figure 7.8. In this system the input is the motion of the support $x(t)$ and the output is the resulting motion of the mass $y(t)$. Physically, this could represent a shock mounting system in which the mass M is to be protected, or isolated, from the motion of its environmental support $x(t)$. To determine the equation of motion of this system we perform a force balance. The forces acting on the mass are those transmitted to

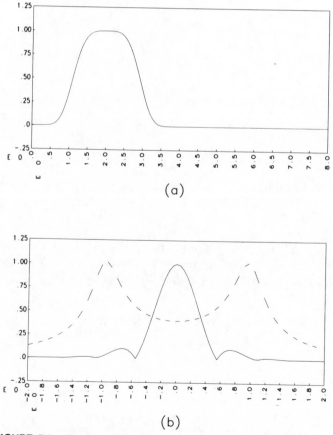

(a)

(b)

FIGURE 7.6. Example problem input function in frequency and time domain.

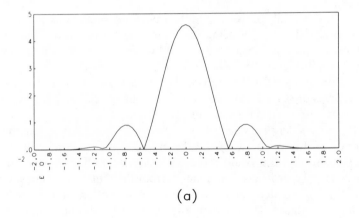

(a)

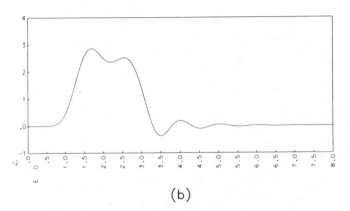

(b)

FIGURE 7.7.　Example problem output and transform.

it by the spring $K[x(t) - y(t)]$ and the damper $Cs[x(t) - y(t)]$. Note that in these expressions $x(t) - y(t)$ is the total compression of the spring and damper. This force balance results in the differential equation

$$(7.17) \qquad \left(s^2 + \frac{sC}{M} + \frac{K}{M}\right) y(t) = \left(\frac{sC}{M} + \frac{K}{M}\right) x(t),$$

which, in turn, results in the transfer function

$$(7.18) \qquad H(w) = \frac{K/M + i2\pi wC/M}{(K/M - 4\pi^2 w^2) + i2\pi wC/M}.$$

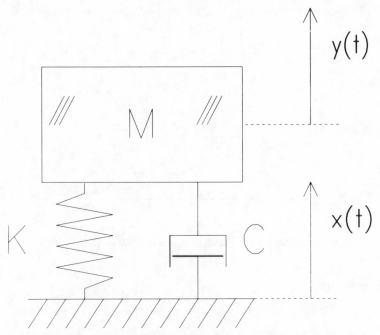

FIGURE 7.8. Mechanical damped oscillator with support excitation.

To make this equation a little more pleasing, we make the following change of variables:

$$(7.19) \qquad w_0 = \frac{\sqrt{K/M}}{2\pi},$$

$$(7.20) \qquad \zeta = \frac{C}{\sqrt{KM}}.$$

where w_0 is the natural frequency of the system and ζ is called the damping ratio. When these substitutions are made in Equation (7.18) (and after a little algebra), we obtain the following normalized version of the transfer function:

$$(7.21) \qquad H(w) = \frac{1 + i\zeta w/w_0}{1 - (w/w_0)^2 + i\zeta w/w_0}$$

The amplitude of this transfer function is known as the transmission factor

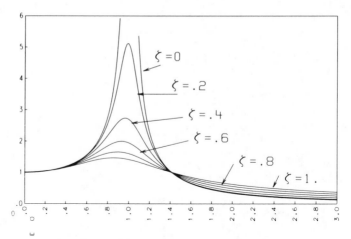

FIGURE 7.9. Transfer function of Equation (7.17).

and is given by

$$(7.22) \qquad |H(w)| = \left\{ \frac{1 + \zeta^2 (w/w_0)^2}{\left[1 - (w/w_0)^2\right]^2 + \zeta^2 (w/w_0)^2} \right\}^{1/2}$$

It is plotted in Figure 7.9 for various values of ζ (i.e., ζ = 0, 0.2, 0.4, 0.6, 0.8, and 1).

ELECTRICAL SYSTEMS

In this section we determine the equations of operation and transfer functions of a resistor, an inductor, and a capacitor, as well as several (series and parallel) combinations of these elements.

From basic electronics we recall that the voltage drop across a resistor $e(t)$ (Figure 7.10a) is given by

$$(7.23) \qquad e(t) = Ri(t),$$

where R is the resistance value and $i(t)$ is the current flowing through the resistor. We may consider the resistor to be a system in which the input is the current $i(t)$ and the output is the resulting voltage drop $e(t)$. Upon comparing Equation (7.23) to Equation (7.1) with $b_0 = 1$, $a_0 = R$, and all other constants

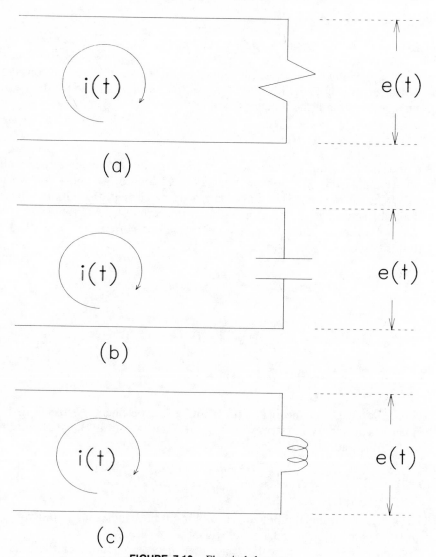

(a)

(b)

(c)

FIGURE 7.10. Electrical elements.

equal to zero, we determine the transfer function of a resistor to be

(7.24) $$H(w) = R.$$

The relationship between the voltage drop across an inductor $e(t)$ and the current flowing through the inductor (Figure 7.10c) is

$$e(t) = L\frac{di(t)}{dt},$$

or

(7.25) $$e(t) = sLi(t),$$

where L is the inductance value (in henrys) of this element. When we consider this element as a system with input $i(t)$ and output $e(t)$ and apply the usual techniques, we obtain the transfer function

(7.26) $$H(w) = 2\pi i w L = 2\pi w L e^{i\pi/2}.$$

We note that this transfer function is complex with amplitude $(2\pi w L)$ and phase $(\pi/2)$. Again, based on the remarks in Chapter 6, we know that when the input to an inductor is a sinusoidally varying current, the output voltage will also be sinusoidal; however, it will lead the current by $\pi/2$ radians.

The voltage drop across a capacitor $e(t)$ due to a current flow $i(t)$ (Figure 7.10b) is described by the equation

$$C\frac{de(t)}{dt} = i(t),$$

or

(7.27) $$sCe(t) = i(t),$$

(7.28) $$e(t) = \frac{i(t)}{sC},$$

where C is the capacitance value (in farads) of this element. As usual, if this element is considered a system with input $i(t)$ and output $e(t)$, we obtain the transfer function

(7.29) $$H(w) = \frac{1}{2\pi i w C} = \frac{1}{2\pi w C}e^{-i\pi/2}.$$

Thus we see a sinusoidal input current will result in a sinusoidal output voltage that lags behind by $\pi/2$ radians.

We now consider electrical circuits that are combinations of these basic elements. Just as we did with mechanical systems, we begin with a classical system, shown in Figure 7.11. In this system we consider the input to be the applied voltage $e_1(t)$ and the output to be the resulting voltage variation across the capacitor $e_2(t)$. To determine the differential equation that describes this circuit we use Kirchhoff's first law, which states that for any loop the summation of the voltage drop across each element is equal to the summation of the applied voltages. In the preceding system this loop analysis results in

$$e_1(t) = Ri(t) + \frac{1}{sC}i(t) + sLi(t).$$

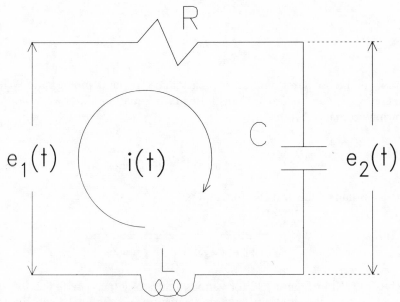

FIGURE 7.11. Electrical damped oscillator system.

The voltage variation across the capacitor is given by

$$e_2(t) = \frac{1}{sC} i(t).$$

Solving the preceding two equations for $e_2(t)$ in terms of $e_1(t)$ we obtain

(7.30) $$(s^2 LC + sRC + 1) e_2(t) = e_1(t).$$

Now, using Equations (7.1), (7.3), (7.5), and (7.6) we determine the transfer function to be

(7.31) $$H(w) = \frac{1}{1 - 4\pi^2 w^2 LC + 2\pi i w CR}.$$

We now make the change of variables

(7.32) $$w_0 = \frac{1}{2\pi \sqrt{LC}},$$

and

$$(7.33) \qquad\qquad\qquad \zeta = R\sqrt{\frac{C}{L}},$$

where w_0 is the oscillator natural frequency and ζ is the damping ratio. When Equations (7.32) and (7.33) are substituted into Equation (7.31) we obtain the following normalized form:

$$(7.34) \qquad\qquad H(w) = \frac{1}{1 - (w/w_0)^2 + i\zeta(w/w_0)}.$$

The magnitude of this transfer function is

$$(7.35) \qquad\qquad |H(w)| = \frac{1}{\sqrt{\left[1 - (w/w_0)^2\right]^2 + \zeta^2(w/w_0)^2}}$$

which is plotted in Figure 7.12 for various values of ζ (i.e., ζ = 0, 0.2, 0.4, 0.6, 0.8, and 1). Note the similarity between this figure and Figure 7.4 which is (the amplitude of) the transfer function of the damped mechanical oscillator of Equation (7.13). These two systems are known as duals of each other. That is, they are two systems that are physically quite different but that share the same type of differential equation. The input voltage $e_1(t)$ of the electrical system is analogous to the driving force $f(t)$ of the mechanical system. Similarly, the output voltage $e_2(t)$ is analogous to the output motion $x(t)$. The resistor serves the same function in the electrical system as the damper in the mechanical

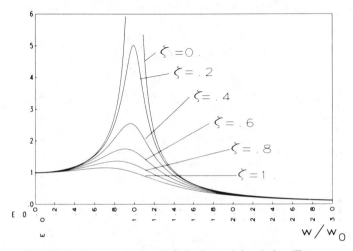

FIGURE 7.12.   Transfer function of a damped electrical oscillator.

system which is the dissipation of energy. In the electrical system the inductor–capacitor combination determines the natural frequency, whereas, in the mechanical system the mass–spring combination does the job.

Next we examine the low pass filter circuit shown in Figure 7.13. In this system the input is voltage $e_1(t)$ and the output is considered to be the resulting voltage across the capacitor $e_2(t)$. Using Kirchhoff's first law for the enclosed loop we obtain the equation

$$Ri(t) + \frac{i(t)}{sC} = e_1(t).$$

Also, the voltage drop across the capacitor is

$$e_2(t) = \frac{i(t)}{sC}.$$

Solving these two equations for $e_2(t)$ in terms of $e_1(t)$ we obtain

(7.36)
$$\left(R + \frac{1}{sC}\right)e_2(t) = \frac{e_1(t)}{sC},$$

which produces the transfer function

(7.37)
$$H(w) = \frac{1}{1 + 2\pi iwRC}.$$

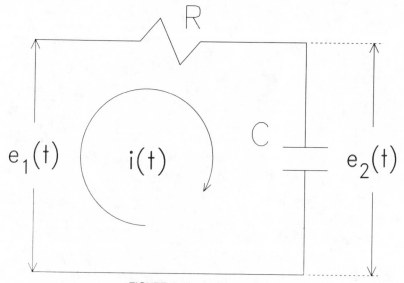

FIGURE 7.13. Low pass filter.

Functions of this form are known as Lorentzian functions. The half-power point of a Lorentzian function is defined as the frequency w_p such that

$$\| H(w_p) \|^2 = \tfrac{1}{2}.$$

For our particular case we have

$$\| H(w_p) \|^2 = \frac{1}{1 + 4\pi^2 w_p^2 R^2 C^2} = \frac{1}{2}$$

and thus we see the half-power point is

(7.38) $$w_p = \frac{1}{2\pi RC}.$$

Shown in Figure 7.14 is a plot of the magnitude of $H(w)$ versus w/w_p.

We recall that the transfer function of this system, as given by Equation (7.37), is the Fourier transform of the impulse response of this system. Thus we determine this system's response to an impulse by finding the inverse Fourier transform of Equation (7.37). Also, we know that

$$\mathcal{F}[e^{-at}] = \frac{1}{a + 2\pi i w}, \qquad a > 0, t > 0.$$

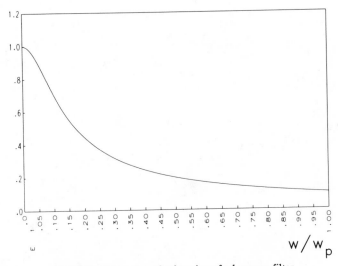

FIGURE 7.14.   Transfer function of a low pass filter.

Thus the impulse response of the low pass electrical filter is

$$\mathscr{F}^{-1}\left(\frac{1}{1 + 2\pi iwRC}\right) = \frac{1}{RC}e^{-t/RC}.$$

Now, if we consider the voltage drop across the resistor in Figure 7.13 as our output, we obtain a high pass filter. This situation is illustrated in Figure 7.15. The differential equations that describe this circuit are

$$Ri(t) + \frac{i(t)}{SC} = e_1(t),$$

$$e_2(t) = Ri(t).$$

Solving the preceding equations for $e_2(t)$ in terms of $e_1(t)$ we obtain

(7.39)
$$\left(R + \frac{1}{sC}\right)e_2(t) = Re_1(t),$$

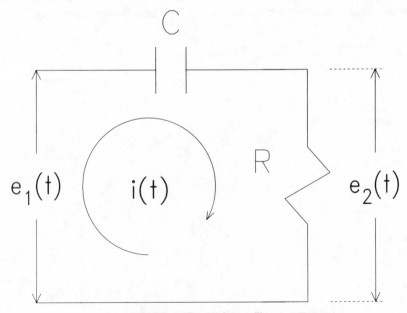

FIGURE 7.15. High pass filter.

which results in the transfer function

(7.40) $$H(w) = \frac{2\pi iwRC}{1 + 2\pi iwRC}.$$

We note that this equation can be written as

$$H(w) = 1 - \frac{1}{1 + 2\pi iwRC},$$

which is simply 1 minus the Lorentzian shaped low pass filter transfer function of Equation (7.37). A little thought reveals that the half-power point of Equation (7.40) is also given by

$$w_p = \frac{1}{2\pi RC}.$$

The magnitude of this transfer function versus w/w_p is shown in Figure 7.16.

As our final illustration in this section, we study the band pass filter system shown in Figure 7.17. Heuristically, we argue that loop 3 offers high impedance to both low frequencies (capacitor C) and high frequencies (inductor L) and relatively low impedance to the natural, or resonant, frequency $w_0 = 2\pi(LC)^{1/2}$. Thus it passes a band of frequencies about w_0 and rejects both the lower and higher frequencies (these rejected frequencies easily pass through the other capacitor and inductor which are connected in parallel. Let us now consider this system more rigorously and write the following loop equations:

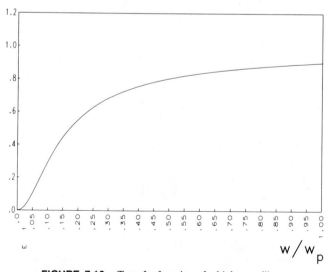

FIGURE 7.16.   Transfer function of a high pass filter.

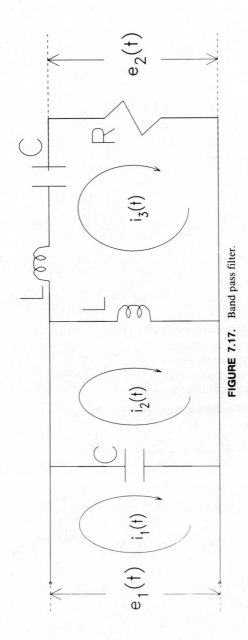

FIGURE 7.17. Band pass filter.

Loop 1

$$\frac{i_1(t)}{sC} - \frac{i_2(t)}{sC} = e_1(t),$$

Loop 2

$$\frac{-i_1(t)}{sC} + \left(\frac{1}{sC} + sL\right)i_2(t) - sLi_3(t) = 0,$$

Loop 3

$$-sLi_2(t) + \left(2sL + R + \frac{1}{sC}\right)i_3(t) = 0.$$

Also, the relationship between the current $i_3(t)$ and the voltage $e_2(t)$ is

$$e_2(t) = Ri_3(t).$$

Solving the preceding equations for $e_2(t)$ in terms of $e_1(t)$ yields

(7.41) $$\left[(sL)^2 + sRL + \frac{L}{C}\right]e_2(t) = sLRe_1(t),$$

and this results in the transfer function

(7.42) $$H(w) = \frac{2\pi iwRC}{1 - 4\pi^2 w^2 LC + 2\pi iwRC}.$$

We now define the center frequency as

$$w_c = \frac{1}{2\pi\sqrt{LC}}$$

and amplitude/width factor ξ as

$$\xi = R\sqrt{C/L}.$$

If these two expressions are used in Equation (7.42), we obtain the following normalized form of the transfer function:

(7.43) $$H(w) = \frac{i\xi(w/w_c)}{1 - (w/w_c)^2 + i\xi(w/w_c)}.$$

The amplitude of this equation is plotted against w/w_c in Figure 7.18 for various values of ξ (i.e., $\xi = 0.01, 0.05, 0.1, 0.2, 0.6, 1$). We note from this figure that the greater the value of ξ (and hence R), the wider the band pass of the filter. We can also see that these curves are not symmetrical about the center frequency w_c. Still, we define the half-power frequencies as those frequencies w_b such that

$$\|H(w_b)\|^2 = \tfrac{1}{2}.$$

Using this definition in Equation (7.43) and a little algebra we can obtain the formula

$$\frac{w_b^2}{w_c^2} = \frac{(2 + \xi^2) +/- \sqrt{(2 + \xi^2)^2 - 4}}{2}.$$

For example, when $\xi = 1$ we find the half-power frequencies to be $0.62w_c$ and $1.62w_c$ ($-1.62w_c$ and $-0.62w_c$ for negative frequencies). The bandwidth of this filter is given as the difference between these two (positive) values.

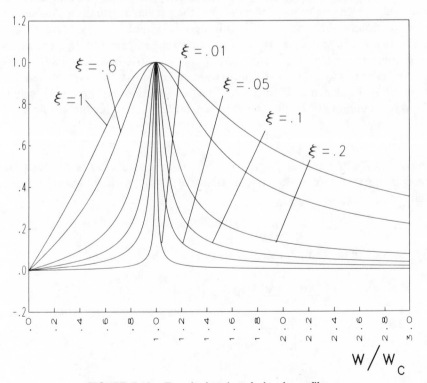

FIGURE 7.18.   Transfer function of a band pass filter.

TRANSVERSE VIBRATION OF AN INFINITE STRING

Up to this point we have studied mechanical systems in which the spring and mass were lumped into individual elements and it was assumed that every point of the mass vibrated in unison. In this section we look at vibrations of a deformable medium in which different locations on the mass are allowed to have different motions. Perhaps the simplest and yet most instructive such system is the transverse motion of a stretched string. Shown in Figure 7.19 is a section of an elastic string of mass ρ per unit length held under constant tension T by equal and oppositely directed forces applied at its ends. In this situation we denote the transverse displacement of the string as $\eta(x, t)$ which is obviously a function of both the position along the string and time. It can quite easily be shown[†] that *for small displacements* $\eta(x, t)$, the governing equation of motion of this string is the one-dimensional scalar wave equation:

$$(7.44) \qquad \frac{\partial^2 \eta(x, t)}{\partial x^2} = \frac{1}{c^2} \frac{\partial^2 \eta(x, t)}{\partial t^2}, \quad \text{where } c = \sqrt{\frac{T}{\rho}}.$$

There are several methods by which a solution to this equation can be obtained. Which one to use depends upon the particular situation. As our first case we assume that the string is infinitely long and initially horizontal, or undisplaced. Then at some arbitrary location x we impart a motion (or signal) with temporal history $f(t)$. That is, our input is strictly a function of time. Inasmuch as this location is arbitrary, we set it equal to 0 for convenience. This situation is illustrated in Figure 7.20. For this situation our method of solving the wave equation (7.44) is to assume a separable function for the solution:

$$(7.45) \qquad \eta(x, t) = \psi(x)\varphi(t),$$

where $\psi(x)$ is the spatial portion of the solution, and $\varphi(t)$ is the temporal portion. Furthermore, we assume a particular form for the temporal portion, namely,

$$(7.46) \qquad \varphi(t) = e^{2\pi i w t}.$$

By methods of elementary calculus we obtain

$$\frac{d\varphi(t)}{dt} = 2\pi i w e^{2\pi i w t},$$

$$\frac{d^2\varphi(t)}{dt^2} = -4\pi^2 w^2 e^{2\pi i w t}.$$

[†]For example, see text by Raven listed in Bibliography at end of this chapter.

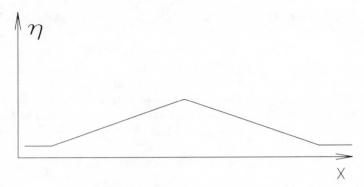

FIGURE 7.19. Portion of an elastic string.

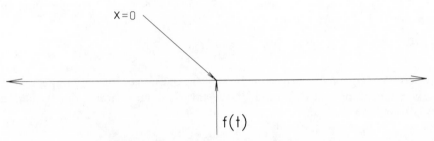

FIGURE 7.20. Infinite string with temporal input.

Therefore [Equation (7.45)],

$$\frac{\partial^2 \eta(x, t)}{\partial t^2} = -4\pi^2 w^2 e^{2\pi iwt}\psi(x),$$

$$\frac{\partial^2 \eta(x, t)}{\partial x^2} = \frac{d^2\psi(x)}{dx^2} e^{2\pi iwt}.$$

When these two equations are substituted back into Equation (7.44) and exp($2\pi iwt$) cancelled, we obtain

(7.47) $$\frac{d^2\psi(x)}{dx^2} + k^2\psi(x) = 0, \quad \text{where } k = \frac{2\pi w}{c}.$$

Equation (7.47) is called the Helmholtz equation in space and its solution is simple and well known:

(7.48) $$\psi(x) = Ae^{ikx} + Be^{-ikx} = Ae^{2\pi iwx/c} + Be^{-2\pi iwx/c},$$

where A and B are constants determined by the boundary conditions of the particular problem. Combining Equations (7.45), (7.46), and (7.48) we find

$$(7.49) \qquad \eta(x, t) = \left(A e^{2\pi i w x / c} + B e^{-2\pi i w x / c} \right) e^{2\pi i w t}.$$

We now consider the string to be a system with temporal input $e(t) = \exp(2\pi i w t)$ and resulting output $\eta(x, t)$. Thus Equation (7.49) tells us that $e(t)$ is an eigenfunction for this system and that its transfer function is

$$H_x(w) = A e^{2\pi i w x / c} + B e^{-2\pi i w x / c},$$

which is valid for any and all locations of x. Using this transfer function we are able to determine the output of the system when the input is an arbitrary function $f(t)$ [with Fourier transform $F(w)$]. That is, in the frequency domain we are able to write

$$Y(x, w) = H_x(w) F(w),$$

where $Y(x, w)$ is the Fourier transform of the output $y(x, t)$ (with respect to t). Therefore, using the definition of the inverse Fourier transform [Equation (3.2)] we have

$$y(x, t) = \int_{-\infty}^{\infty} H_x(w) F(w) e^{2\pi i w t} \, dw,$$

or

$$y(x, t) = A \int_{-\infty}^{\infty} F(w) e^{2\pi i w x / c} e^{2\pi i w t} \, dw$$

$$+ B \int_{-\infty}^{\infty} F(w) e^{-2\pi i w x / c} e^{2\pi i w t} \, dw.$$

Direct application of the first shifting theorem 3.2 yields the result

$$(7.50) \qquad y(x, t) = A f\left(t + \frac{x}{c} \right) + B f\left(t - \frac{x}{c} \right).$$

Equation (7.50) is the so-called D'Alembert's solution to the wave equation. This equation tells us that the input function $f(t)$ sets up two traveling waves on the string propagating in opposite directions. Furthermore, these traveling waves have the same spatial shape as the temporal history of the input function. Shown in Figure 7.21 is the time history of a forcing function that is

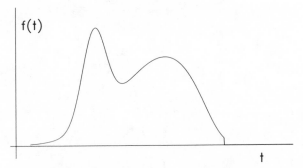

FIGURE 7.21.   Temporal history of forcing function.

injected into the string at location $x = 0$. Figure 7.22a, b, and c shows a "snapshot" of the string at successive times $t_1 < t_2 < t_3$.

We now look at this problem from a slightly different point of view. Let us consider the string to be a system in which the input is strictly a function of space $f(x)$. Again we attack the problem by assuming that the wave equation solution is a separable function as per Equation (7.45). This time, however, we assume a particular form for the spatial portion, namely,

$$(7.51) \qquad\qquad \psi(x) = e^{2\pi i \nu x}.$$

Using methods analogous to those of our previous development, we find that this assumption, when substituted into the wave equation, results in the Helmholtz equation in time:

$$(7.52) \qquad \frac{d^2\varphi(t)}{dt^2} + l^2\varphi(t) = 0, \quad \text{where } l = 2\pi\nu c.$$

When this equation is solved and the results substituted into Equation (7.45) we obtain

$$(7.53) \qquad \eta(x, t) = \psi(x)\varphi(t) = \left(Ce^{2\pi i\nu ct} + De^{-2\pi i\nu ct}\right)e^{2\pi i\nu x}.$$

Inasmuch as we are now considering the string to be a system with a spatial function as input, the preceding equation tells us that the transfer function of this system is

$$H_t(\nu) = Ce^{2\pi i\nu ct} + De^{-2\pi i\nu ct},$$

which is valid for any and all times t.

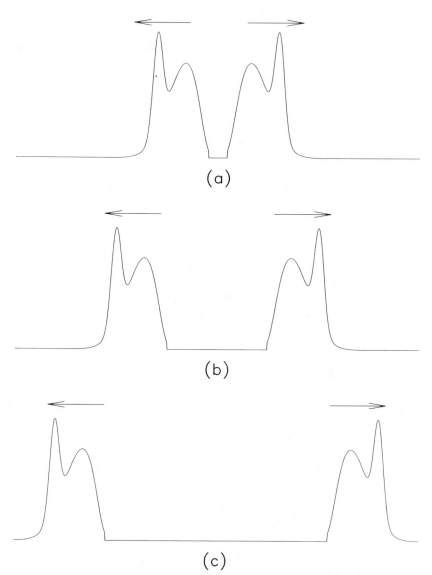

FIGURE 7.22.   Traveling waves resulting from initial temporal displacement.

We now determine the response of this system to an arbitrary spatial input function $f(x)$ [with Fourier transform $F(\nu)$]. In the frequency domain the transform of the resulting output $Y(\nu, t)$ is simply the product of the transfer function $H_t(\nu)$ and $F(\nu)$:

$$Y(\nu, t) = H_t(\nu)F(\nu) \quad \text{or} \quad Y(\nu, t) = \left(Ce^{2\pi i\nu ct} + De^{-2\pi i\nu ct}\right)F(\nu).$$

In the time domain this becomes

$$y(x, t) = C\int_{-\infty}^{\infty} F(\nu)e^{2\pi i\nu ct}e^{2\pi i\nu x}\, d\nu + D\int_{-\infty}^{\infty} F(\nu)e^{-2\pi i\nu ct}e^{2\pi i\nu x}\, d\nu.$$

Direct application of the first shifting Theorem 3.2 yields the result:

(7.54) $$y(x, t) = Cf(x + ct) + Df(x - ct).$$

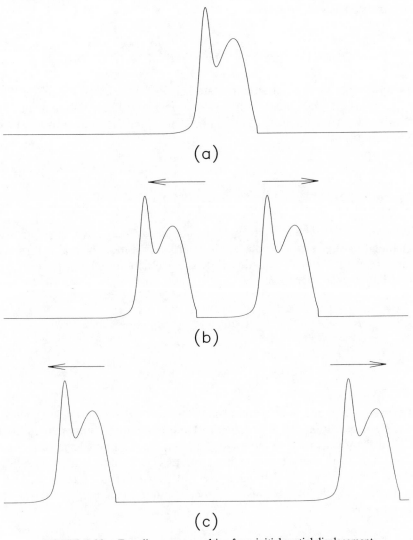

FIGURE 7.23.   Traveling waves resulting from initial spatial displacement.

This is simply another form of D'Alembert's solution of Equation (7.50). This equation tells us that if an arbitrary shape $f(x)$ is impressed upon the string at any time t, then this will result in two traveling waves that will retain this shape while propagating (in opposite directions). For example, shown in Figure 7.23a is a string which is initially at rest [i.e., $dy(x,0)/dt = 0$] and at time $t = 0$ has the arbitrary shape $f(x)$ impressed upon it. Figures 7.23b and 7.23c show this same string at later times t_1 and t_2 ($t_1 < t_2$). These "snapshots" should be compared to those of Figures 7.22 in which a temporal forcing function was used to generate the waves. The spatially generated waves (Figure 7.23) have the same shape, whereas the temporally generated ones (Figure 7.22) are mirror images of each other.

TRANSVERSE VIBRATION OF A FINITE STRING

We now come to the last situation studied in this chapter, namely, a finite string with both ends fixed. This is shown in Figure 7.24. We point out here that since both Equations (7.49) and (7.53) are solutions to the wave equation, then so too is their sum

(7.55)

$$\eta(x, t) = \left(Ae^{2\pi iwx/c} + Be^{-2\pi iwx/c} \right)e^{2\pi iwt} + \left(Ce^{2\pi i\nu ct} + De^{-2\pi i\nu ct} \right)e^{2\pi i\nu x}.$$

This can be verified by substituting Equation (7.55) back into the wave equation (7.44). We wish to place this equation in a more workable form and, therefore, we use Euler's equation [Equations (1.22) and (1.23)] to rewrite the preceding expression as

$$(7.56) \quad \eta(x, t) = a \sin \frac{2\pi wx}{c} \sin 2\pi wt + b \sin \frac{2\pi wx}{c} \cos 2\pi wt$$

$$+ c \cos \frac{2\pi wx}{c} \sin 2\pi wt + d \cos \frac{2\pi wx}{c} \cos 2\pi wt.$$

The constants a, b, c, and d are to be determined by the particular boundary and initial conditions of the problem. If there are any doubts that Equation (7.56) is truly a solution to the wave equation, they can be alleviated by direct substitution of this equation into Equation (7.44). We now apply this solution to the finite string problem and use the prevailing boundary and initial conditions to solve for the constants a, b, c, and d. Since $\eta(0, t) = 0$, for all t, we must have

$$c \sin 2\pi wt + d \cos 2\pi wt = 0.$$

FIGURE 7.24. Finite string with fixed ends.

The only way that this equation can be satisfied for all t is if $c = d = 0$. Equation (7.56) now becomes

$$\eta(x, t) = \sin \frac{2\pi w x}{c} \left(a \sin 2\pi w t + b \cos 2\pi w t \right).$$

Let us now assume (as an initial condition) that the string is initially at rest;

$$\frac{d\eta(x, 0)}{dt} = 0.$$

To apply this condition we first determine

$$\frac{d\eta(x, t)}{dt} = 2\pi w \sin \frac{2\pi w x}{c} \left(a \cos 2\pi w t - b \sin 2\pi w t \right).$$

For $t = 0$ we have

$$\frac{d\eta(x, 0)}{dt} = 0 = a 2\pi w \sin \frac{2\pi w x}{c}$$

and thus $a = 0$ and we arrive at

$$\eta(x, t) = b \sin \frac{2\pi w x}{c} \cos 2\pi w t.$$

We now apply the boundary condition that the string must remain fixed at $x = L$:

$$\eta(L, t) = 0 = b \sin \frac{2\pi w L}{c} \cos 2\pi w t \quad \text{for all } t.$$

The trivial solution is $b = 0$; however, we are interested in the nontrivial solution and, therefore,

$$\sin \frac{2\pi wL}{c} = 0,$$

(7.57) $$w = \frac{nc}{2L} \qquad n = 1, 2, \ldots .$$

Thus we see that we have an infinite number of solutions

$$\eta(x, t) = b_n \sin \frac{\pi nx}{L} \cos \frac{\pi nct}{L} \qquad n = 1, 2, \ldots ,$$

or

(7.58) $$\eta(x, t) = b_n \sin 2\pi k_n x \cos 2\pi w_n t.$$

We call $k_n = n/2L$ the spatial frequency and $w_n = nc/2L$ the temporal frequency. The constants b_n are determined by the initial displacement of the string; that is, $\eta(x, 0) = b_n \sin 2\pi k_n x$.

Thus we see that both the spatial and temporal frequencies are multiples of basic frequencies $1/2L$ and $c/2L$, respectively. It is also clear that these frequencies are not independent. That is, if the spatial frequency is k, then the temporal frequency w must be equal to ck. The basic spatial frequency is determined solely by the length of the string, whereas the temporal frequency depends upon the length, tension, and mass per unit length of the string:

$$\frac{c}{2L} = \frac{1}{2L} \sqrt{\frac{T}{\rho}} .$$

An excellent example of this type of system is a fixed string on a musical instrument such as a guitar. The temporal frequency with which the string vibrates determines the sound that it produces. The pitch of the string is, for the most part, determined by the basic temporal frequency $c/2L$. Thus the pitch is increased as the string is shortened and/or tightened. Also the thinner (less density per unit length) the string—the higher the pitch.

Inasmuch as Equation (7.58) is a solution to the wave equation for any and all values of n, the total solution is obtained by adding these together:

(7.59) $$y(x, t) = \sum_{n=1}^{\infty} b_n \sin \frac{\pi nx}{L} \cos \frac{\pi nct}{L} .$$

Let us now assume that the string has an initial shape given by

$$y(x, 0) = g(x),$$

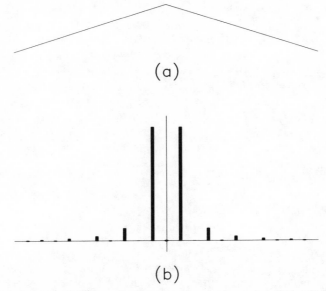

(a)

(b)

FIGURE 7.25. Guitar string plucked in middle and resulting frequency content.

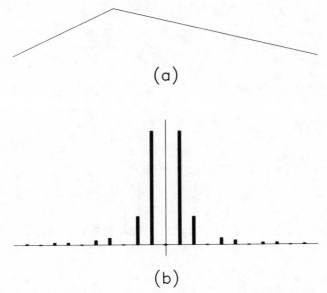

(a)

(b)

FIGURE 7.26. Guitar string plucked at $L/3$ and resulting frequency content.

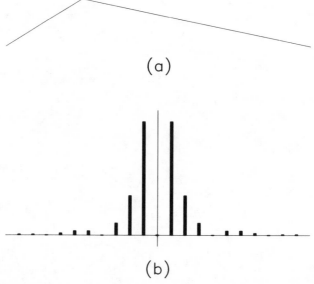

FIGURE 7.27. Guitar string plucked at $L/4$ and resulting frequency content.

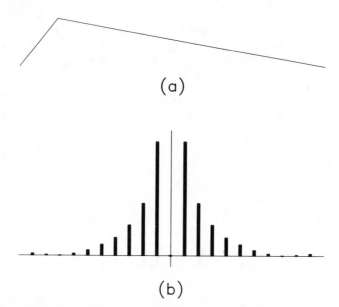

FIGURE 7.28. Guitar string plucked at $L/8$ and resulting frequency content.

or

$$g(x) = \sum_{n=1}^{\infty} b_n \sin \frac{2\pi nx}{2L}.$$

Therefore, we see that the constants b_n are determined by the initial shape $g(x)$. In fact, they are the Fourier series coefficients of the (odd) function $g(x)^{\dagger}$ and are given by (Chapter 2)

$$b_n = \frac{2}{L} \int_0^L g(x) \sin \frac{2\pi nx}{2L} dx.$$

The constant b_n is a measure of the spatial frequency k_n present in the vibrational motion of the string and, as we have previously noted, it is also a measure of the temporal frequency w_n present. Therefore, the initial shape of the string completely determines both the spatial and temporal frequencies with which the string will vibrate.

Let us return to the guitar string illustration and show that the "sound" the string produces is determined by its initial shape, or in other words, how it is "plucked." We must first clarify what we mean by sound, or more properly, timbre. By timbre is meant those features of a musical sound that give an instrument (or voice) its distinctive characteristic, even when the pitch (basic frequency) and loudness (amplitude) are the same. The timbre is a qualitative measure of an instrument but it is determined mostly by the frequency content of the sound. Let us first assume that we "pluck" the guitar string in the center, which results in the initial displacement shown in Figure 7.25a. Evaluation of the Fourier series coefficients b_n for this situation results in the normalized frequency spectrum (C_j versus j) shown in Figure 7.25b. Shown in Figures 7.26–7.28 is the frequency content of the string as it is "plucked" at various locations nearer to the end. Note the shift in the frequency content or the increase in strength of the higher frequencies (or overtones).

SUMMARY

In this chapter we have studied both mechanical and electrical vibrational systems. We examined the governing equations of each of the individual elements and then showed how to determine the differential equation(s) that governed combinations of these elements. We then showed how to determine the transfer function (and, therefore, the impulse response) from this govern-

---

$\dagger$ Note that for our analysis we consider the function to be periodic over the interval $[-L, L]$ with $g(-x) = -g(x)$. Physically, however, we are only interested in the solution over the half-interval $[0, L]$.

ing, or differential, equation. Obviously, it is not possible to consider all the interesting examples in just one chapter, but those that were covered should provide enough insight into the methods used to enable the reader to tackle most any linear mechanical and/or electrical system. Also in this chapter we presented a treatment of the wave equation in one dimension as it related to the transverse motion of a vibrating string. We discussed several specific and, we hope, interesting situations. This presentation also serves as an introduction to the solution of the wave equation in three dimensions that we must study in the next chapter when we consider optical applications of Fourier analysis.

BIBLIOGRAPHY

Raven, F. H., *Mathematics of Engineering Systems*, McGraw-Hill, New York, 1966.

Carson, J. R., *Electric Circuit Theory and Operational Calculus*, McGraw-Hill, New York, 1926.

Koenig, H. E., *Electromechanical Systems Theory*, McGraw-Hill, New York, 1961.

CHAPTER

8

OPTICS

Optics, in simple terms, is the study of propagation of light from one spatial plane to another. Often objects such as lenses and/or apertures are placed between these planes to modify the light in some desired way. The Fresnel diffraction equation provides a mathematical description of this propagation phenomenon and, as it turns out, this Fresnel equation can be written in terms of Fourier transforms. This fortunate fact permits us to use our knowledge of Fourier analysis to both solve and gain valuable insight into many common optical problems. In this chapter we first give a nonrigorous discussion of the derivation of the Fresnel equation from Maxwell's equations. We then proceed to discuss various topics in optics using this and our Fourier analysis background.

THE FRESNEL DIFFRACTION EQUATION

The Fresnel diffraction equation, as already mentioned, provides us with a mathematical description of how optical radiation propagates from one plane in space to another. Under "proper conditions" this equation turns out to be extremely accurate. In this section we examine what these "proper conditions" are. We do so by a heuristic derivation of this equation from Maxwell's equations. This presentation is not intended to be rigorous but instead is included to point out the various limitations of the Fresnel equation.

The optical spectrum is a subset of the much larger electromagnetic spectrum which ranges from low frequency radio waves to the very high frequency cosmic rays. In between these extremes fall radar, microwaves, the optical spectrum (which includes infrared, visible, and ultraviolet radiation), X-rays, and gamma rays. All these radiations have in common the fact that they propagate through space as transverse electromagnetic waves and, in a vacuum, their speed of propagation c is the product of their frequency ν and their wavelength λ (i.e., $\lambda\nu = c$). c is called the speed of light and is equal to 3×10^{10} cm/sec.

The main difference between these radiations is the way in which they are produced. Radio waves, for example, are generated by electrons oscillating in an undamped LC circuit and transferring their energy back and forth between the electric field of the capacitor and the magnetic field of the inductor. Infrared radiation is produced by the electric and magnetic fields set up in vibrating and rotating molecules. The visible and ultraviolet spectrum is generated by electrons performing transitions across atomic fields. Nuclear transitions are responsible for X-rays and gamma rays.

The mathematical description of how electromagnetic radiation behaves in the presence of matter is provided by Maxwell's equations which are, in fact, simply a collection of other well-known equations. These are given in Table 8.1. In these equations $\mathbf{E}(x; t)$ and $\mathbf{H}(x; t)$ are the electric and magnetic field vectors, respectively. Both are vector functions of space $[\mathbf{x} = (x, y, z)]$ and time. Electron (or charged particle) conditions in the material are given by ρ the charge density and $\mathbf{J}(x; t)$ the current density. Properties of the material are specified by ε the permittivity, μ the permeability, and σ the conductivity. Of interest to us is the situation where the propagation medium is free space (or a dielectric) for which $\varepsilon = \mu = 1$ (in electrostatic units) and $\rho = \sigma = 0$. When these assumptions are substituted into Maxwell's equations we obtain

(8.1)
$$\mathbf{J}(\mathbf{x}; t) = 0,$$

(8.2)
$$\nabla \times \mathbf{E}(\mathbf{x}; t) = -\frac{1}{c}\frac{\partial \mathbf{H}(\mathbf{x}; t)}{\partial t}.$$

(8.3)
$$\nabla \times \mathbf{H}(\mathbf{x}; t) = \frac{1}{c}\frac{\partial \mathbf{E}(\mathbf{x}; t)}{\partial t}.$$

(8.4)
$$\nabla \cdot \mathbf{E}(\mathbf{x}; t) = 0.$$

(8.5)
$$\nabla \cdot \mathbf{H}(\mathbf{x}; t) = 0.$$

Using the vector identity

$$\nabla \times (\nabla \times \mathbf{A}) = \nabla(\nabla \cdot \mathbf{A}) - \nabla^2\mathbf{A},$$

TABLE 8.1.   Maxwell's equations

(ME-1) *Faraday*

$$\nabla \times \mathbf{E}(\mathbf{x}; t) = -\frac{\mu}{c}\frac{\partial \mathbf{H}(\mathbf{x}; t)}{\partial t}$$

(ME-2) *Ampere*

$$\nabla \times \mathbf{H}(\mathbf{x}; t) = \frac{\varepsilon}{c}\frac{\partial \mathbf{E}(\mathbf{x}; t)}{\partial t} + \frac{\mathbf{J}(\mathbf{x}; t)}{c}$$

(ME-3) *Gauss (Electric Field)*

$$\nabla \cdot \mathbf{E}(\mathbf{x}; t) = \rho$$

(ME-4) *Gauss (Magnetic Field)*

$$\nabla \cdot \mathbf{H}(\mathbf{x}; t) = 0$$

(ME-5) *Ohm*

$$\mathbf{J}(\mathbf{x}; t) = \sigma \mathbf{E}(\mathbf{x}; t)$$

we can uncouple Equations (8.2) and (8.3). To do so we first form the cross product of both sides of Equation (8.2) with ∇:

$$\nabla \times (\nabla \times \mathbf{E}) = \nabla \times \left(-\frac{1}{c}\frac{\partial \mathbf{H}}{\partial t}\right).$$

Using the previously mentioned identity, and also assuming that we can interchange the order of differentiation on the right-hand side, we obtain

$$\nabla(\nabla \cdot \mathbf{E}) - \nabla^2 \mathbf{E} = -\frac{1}{c}\frac{\partial}{\partial t}(\nabla \times \mathbf{H}).$$

Now using Equations (8.3) and (8.4) we find

$$-\nabla^2 \mathbf{E} = -\frac{1}{c}\frac{\partial}{\partial t}\left(\frac{1}{c}\frac{\partial \mathbf{E}}{\partial t}\right) \quad \text{or} \quad \nabla^2 \mathbf{E} = \frac{1}{c^2}\frac{\partial^2 \mathbf{E}}{\partial t^2}.$$

A similar equation holds for the magnetic field vector:

$$\nabla^2 \mathbf{H} = \frac{1}{c^2}\frac{\partial^2 \mathbf{H}}{\partial t^2}.$$

Using additional vector analysis for this case, it is also possible to demonstrate that these two vectors **E** and **H** are transverse to each other and, furthermore, they have the same phase and magnitude. Thus knowledge of either vector is sufficient to completely describe the propagation problem. Finally, inasmuch as these vectors are always perpendicular and propagate in a single direction, we can always choose our coordinate system so that **E** and/or **H** will remain in one plane. That is, they can be treated as scalar functions of time and space.

In summary, for a dielectric such as free space ($\sigma = \rho = 0$ and $\varepsilon = \mu = 1$), Maxwell's equations can be written as two scalar wave equations. Either of these equations describe the propagation of electromagnetic radiation through space. For convenience we define the optical disturbance function $V(\mathbf{x}; t)$ to be either the electric or magnetic component. This function must obviously satisfy the scalar wave equation in three dimensions:

$$(8.6) \qquad \nabla^2 V(\mathbf{x}; t) = \frac{1}{c^2} \frac{\partial^2 V(\mathbf{x}; t)}{\partial t^2}.$$

From here on we limit our attention to radiation within the optical spectrum. Physically, we note that optical detectors (in particular the human eye in the visible spectrum) respond to the modulus squared of $V(\mathbf{x}; t)$. That is to say, they "see" $\|V(\mathbf{x}; t)\|^2$ but not $V(\mathbf{x}; t)$.

Our approach to solving Equation (8.6) is similar to the one we used in Chapter 7 for the one-dimensional wave equation (7.44). In other words, we assume that the solution is a separable function of space and time:

$$(8.7) \qquad V(\mathbf{x}; t) = \psi(\mathbf{x})\varphi(t).$$

Furthermore, we assume a particular form for the temporal solution, namely,

$$(8.8) \qquad \varphi(t) = e^{2\pi i \nu t}.$$

Equation (8.8) is known as the monochromatic assumption. That is, we assume that the temporal behavior of the optical disturbance function is single frequency or single color. When this assumption, along with Equation (8.7), is substituted back into Equation (8.6), we obtain the Helmholtz equation in three variables:

$$(8.9) \qquad \nabla^2 \psi(\mathbf{x}) + k^2 \psi(\mathbf{x}) = 0,$$

or

$$\frac{\partial^2 \psi(x, y, z)}{\partial x^2} + \frac{\partial^2 \psi(x, y, z)}{\partial y^2} + \frac{\partial^2 \psi(x, y, z)}{\partial z^2} + k^2 \psi(x, y, z) = 0,$$

where $k = 2\pi v/c = 2\pi/\lambda$. The task before us is to solve this equation for $\psi(\mathbf{x})$. Just as with the one-dimensional case (Chapter 7), there are several approaches that can be taken depending upon the actual geometry involved. For example, a spherical wave emanating from a point source at $x = 0$ is described by

$$\psi_{sp}(\mathbf{x}) = \frac{e^{ikr}}{r}, \qquad r > 0,$$

where $r^2 = x^2 + y^2 + z^2$ is the distance from the point source to the point of observation. With a reasonable amount of mathematical fortitude it can be shown, by direct substitution into Equation (8.9), that $\psi_{sp}(\mathbf{x})$ is a solution to the Helmholtz equation. Of interest to us here is the boundary value problem of solving Equation (8.9) between two planes as shown in Figure 8.1. Green's theorem provides the mathematical sophistication necessary to solve this problem. However, much greater insight into the physical factors of the problem can be obtained by using the less rigorous Huygen's solution.

The Huygen–Fresnel principle may be stated as: *a geometrical point source of light will give rise to a spherical wave propagating equally in all directions.*

We also note that the Helmholtz equation is a linear differential equation and, therefore, we can use superposition of solutions. That is, if $\psi_1(\mathbf{x})$ and $\psi_2(\mathbf{x})$ are both solutions to the equation, then so too is their sum $\psi(\mathbf{x}) = \psi_1(\mathbf{x}) + \psi_2(\mathbf{x})$.

Let us choose our coordinate system so that the origin is located in plane 1. That is, $z = 0$ in this plane and, therefore, the optical disturbance function at any point in plane 1 is described by $\psi_1(\xi, \eta) = \psi(\xi, \eta, 0)$. $\psi_1(\xi, \eta)$ is not normally defined over the entire plane but rather over some finite region R in

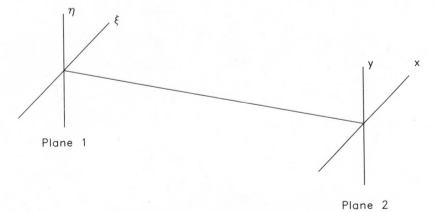

FIGURE 8.1.   Light propagation between two plane.

the plane that physically represents an aperture (see Figure 8.2). Next we partition this region R into a network of n rectangular subregions, one of which is shown in Figure 8.2. This i th subregion is centered at the point (ξ_i, η_i) and has an area of $\Delta_i \xi \Delta_i \eta$. Let us now assume that the value of $\psi_1(\xi, \eta)$ over this entire subregion can be approximated by its value at (ξ_i, η_i). In other words, we approximate the optical disturbance function over the subregion by a point source at (ξ_i, η_i) whose strength is given by $\psi_1(\xi_i, \eta_i)\Delta_i \xi \Delta_i \eta$. Thus the resulting optical disturbance function at any location (x, y, z) in space is given by (see Figure 8.3)

$$\psi_i(x, y, z) = \psi_1(\xi_i, \eta_i)\Delta_i \xi \Delta_i \eta \frac{e^{ikr_i}}{r_i},$$

where r_i is the distance from $(\xi_i, \eta_i, 0)$ to the point (x, y, z).

Although this logic is all very pleasing, it does neglect the important fact that the propagating wave has a preferred direction. Therefore, to account for this we must include an inclination factor $K(\theta_i)$. Although we do not do so here, it can be shown that this factor is given by

$$K(\theta_i) = \frac{-i \cos \theta_i}{\lambda},$$

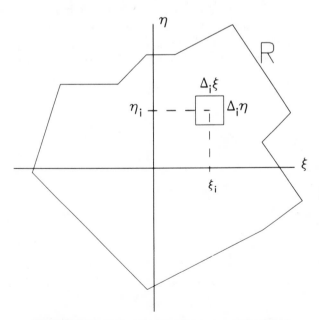

FIGURE 8.2. Region R and partition element in plane 1.

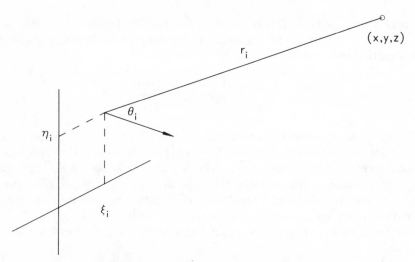

FIGURE 8.3.   Spherical wave emanating from (ξ_i, η_i).

where θ_i is the angle between the normal to plane 1 and the line from $(\xi_i, \eta_i, 0)$ to (x, y, z). Including this inclination factor, the equation becomes

$$\psi_i(x, y, z) = \psi_1(\xi_i, \eta_i) K(\theta_i) \frac{e^{ikr_i}}{r_i} \Delta_i \xi \Delta_i \eta.$$

We repeat this process for each and every subregion and, therefore, based on our previous remarks concerning superposition, the total optical disturbance at the location (x, y, z) is simply the sum of all the point sources over the region R:

$$\psi(x, y, z) = \sum_{i=1}^{n} \psi_1(\xi_i, \eta_i) K(\theta_i) \frac{e^{ikr_i}}{r_i} \Delta_i \xi \Delta_i \eta.$$

If we shrink the size of the rectangular subregions to zero (and thus increase their number to infinity), we know from intermediate calculus that the preceding summation becomes the following double integral:[†]

$$(8.10) \qquad \psi(x, y, z) = \iint_R \psi_1(\xi, \eta) K(\xi, x) \frac{e^{ikr(\xi, x)}}{r(\xi, x)} d\xi \, d\eta.$$

This equation defines the optical disturbance function, at any point in space, resulting from a disturbance in plane 1. We now limit our attention to the situation where the observation point lies in a second plane located a

---

[†] Note that since the inclination angle θ_i depends upon the coordinates (x, y) and (ξ, η), we write the inclination factor K as a function of these coordinates.

distance z from plane 1. This situation is illustrated in Figure 8.4. In this case we write $\psi(x, y, z)$ as $\psi_2(x, y)$ and thus Equation (8.10) becomes the well-known diffraction equation

$$(8.11) \qquad \psi_2(x, y) = \iint_R \psi_1(\xi, \eta) K(\xi, x) \frac{e^{ikr(\xi, x)}}{r(\xi, x)} d\xi \, d\eta.$$

This integral equation provides a very accurate description of the propagation of optical radiation (in free space) from plane 1 to plane 2. However, in its present form it is very difficult to use. Therefore, we make several assumptions, or approximations, that simplify this equation to the Fresnel diffraction equation. These assumptions should be carefully noted because they define the conditions under which the Fresnel equation can be used accurately.

We begin by writing expressions for r and $\cos\theta$. From Figures 8.3 and 8.4 we can see

$$(8.12) \qquad r = \sqrt{(x - \xi)^2 + (y - \eta)^2 + z^2} \,,$$

$$(8.13) \qquad \cos\theta = \frac{z}{r} = \frac{z}{\sqrt{(x - \xi)^2 + (y - \eta)^2 + z^2}}$$

We recall here that the point (ξ, η) is confined to the finite aperture region R in plane 1. We similarly limit our observations in plane 2 to points (x, y)

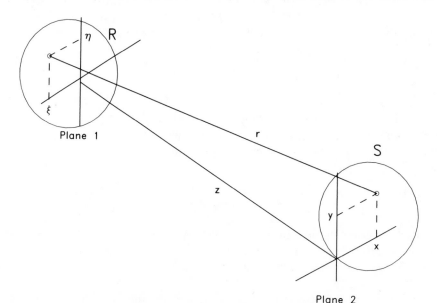

FIGURE 8.4. Propagation between two planes.

within some finite region S (Figure 8.4). Therefore, $(x - \xi)^2 + (y - \eta)^2$ will be bounded by some number a^2, where a is obviously dependent upon some characteristic dimension of both R and S. For example, if R and S are both circular regions of radius r_a, then the bound a^2 is given by

$$a^2 = \left(2r_a\right)^2 + \left(2r_a\right)^2 = 8r_a^2.$$

Now let us consider Equation (8.13) which we first normalize by z:

$$\frac{1}{\sqrt{\left[(x - \xi)^2 + (y - \eta)^2\right]/z^2 + 1}} \geqslant \frac{1}{\sqrt{1 + (a/z)^2}},$$

where (a/z) is the ratio of the (maximum) dimensions in the planes to the distance between the planes. Shown in Figure 8.5 is a plot of $(\cos \theta)_{max}$ versus a/z (solid line). From this plot we see that, for reasonably small values of a/z, $\cos \theta$ is essentially a constant and equal to one. Thus we have

(8.14) $$K(\theta) = \frac{-i \cos \theta}{\lambda} \approx \frac{-i}{\lambda} = K.$$

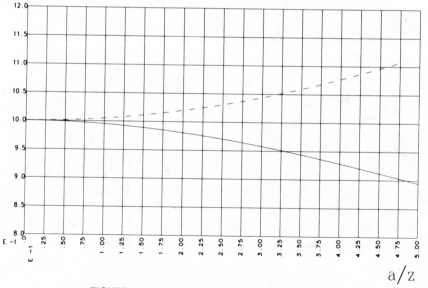

a/z

FIGURE 8.5. Accuracy of small angle approximations.

Now let us rewrite Equation (8.12) as the ratio

(8.15) $$\frac{r}{z} = \sqrt{\frac{(x-\xi)^2}{z^2} + \frac{(y-\eta)^2}{z^2} + 1} \leqslant \sqrt{\left(\frac{a}{z}\right)^2 + 1}$$

This ratio is plotted versus a/z in Figure 8.5 as the dashed line. We see from this figure that for reasonably small values of a/z we can let $r \approx z$ which is independent of ξ and η and, therefore, a constant that can be removed from under the integral sign. The conditions expressed by Equations (8.14) and (8.15) are known as the small angle, or paraxial, approximations. When they are used in Equation (8.11) we find

(8.16) $$\psi_2(x, y) = \frac{K}{z} \iint_R \psi_1(\xi, \eta) e^{ikr(\xi, x)} \, d\xi \, d\eta.$$

Note that in this equation we still have $r(\xi, x)$ in the exponent term. In other words, we did not replace it by z as we did for the $r(\xi, x)$ in the denominator [Equation (8.11)]. The reason for this is that it is multiplied by $k = 2\pi/\lambda$ which, in the optical spectrum, is a rather large number (on the order of 10^3–10^7 cm^{-1}). Therefore, we need a more accurate approximation. We obtain this by rewriting Equation (8.12) as

$$r = z\sqrt{\left(\frac{x-\xi}{z}\right)^2 + \left(\frac{y-\eta}{z}\right)^2 + 1}\,.$$

Using the binomial approximation $\sqrt{1 + \alpha} \approx 1 + \alpha/2$ in the preceding equation we obtain

(8.17) $$r \approx z\left[1 + \frac{(x-\xi)^2}{2z^2} + \frac{(y-\eta)^2}{2z^2}\right] \leqslant z\left(1 + \frac{a^2}{2z^2}\right).$$

Shown in Figure 8.6 is a plot of Equation (8.17) (normalized by the true value of r) versus a/z from which we can see the relative accuracy of this approximation. It is interesting to note that when we approximate Equation (8.12) by Equation (8.17) we are, in effect, approximating a spheroid with an ellipsoid. This is illustrated (in one dimension) in Figure 8.7.

Finally, if we use the approximation of Equation (8.17) in Equation (8.16) we obtain the first form of the Fresnel diffraction equation:

(8.18) $$\psi_2(x, y) = \frac{Ke^{ikz}}{z} \iint_R \psi_1(\xi, \eta) e^{(ik/2z)[(x-\xi)^2 + (y-\eta)^2]} \, d\xi \, d\eta.$$

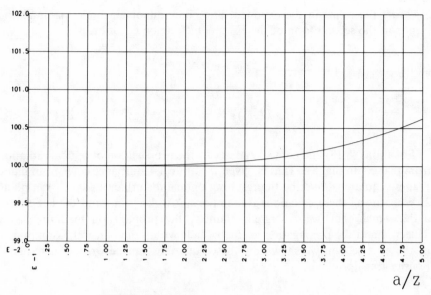

FIGURE 8.6. Relative accuracy of Equation (8.17).

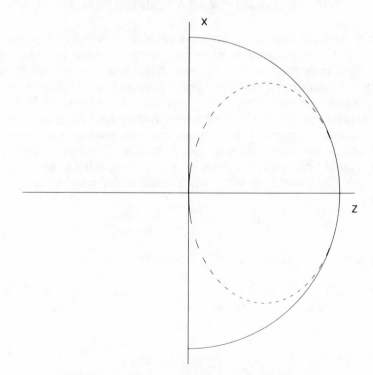

FIGURE 8.7. Parabolic approximation to spherical wave.

205

Squaring the terms in the exponent and some minor algebraic manipulation results in the second form of the Fresnel diffraction equation

$$(8.19) \quad \psi_2(x, y) = \frac{Ke^{ikz}}{z} e^{(ik/2z)(x^2+y^2)}$$

$$\times \iint_R \psi_1(\xi, \eta) e^{(ik/2z)(\xi^2+\eta^2)} e^{-(2\pi i/\lambda z)(x\xi+y\eta)} \, d\xi \, d\eta.$$

Equations (8.18) and/or (8.19) describe the propagation of optical radiation from plane 1 to plane 2. That is, given the optical disturbance $\psi_1(\xi, \eta)$ in plane 1, these equations describe the resulting optical disturbance $\psi_2(x, y)$ in plane 2. When using these equations it is important to keep in mind the conditions under which they were derived. Namely, the propagation medium was a dielectric and the propagation geometry falls within the paraxial assumptions. In the next section we show how these equations can be written in terms of Fourier transforms.

FRESNEL EQUATION AS A FOURIER TRANSFORM

In the previous section we reduced the problem of solving the scalar wave equation to that of integrating the Fresnel diffraction equation (8.18) and/or (8.19). This integration is usually quite difficult and normally cannot be performed analytically. Thus some form of numerical algorithm must be used. In this section we show how both equations can be rewritten in terms of Fourier transforms for which we have already developed the required numerical techniques in Chapter 5. Also, when these equations appear in this form, we are able to use our background and experience in Fourier analysis to gain valuable insight into many problems without actually solving them.

We begin by introducing the aperture function $A(\xi, \eta)$ which is defined as

$$A(\xi, \eta) = \begin{cases} 1, & (\xi, \eta) \varepsilon R, \\ 0, & \text{otherwise.} \end{cases}$$

Using this we are able to rewrite Equation (8.19) with new limits:

$$(8.20) \quad \psi_2(x, y) = C(\mathbf{x}) \iint_{-\infty}^{\infty} g(\xi, \eta) e^{-2\pi i(\xi x/\lambda z + \eta y/\lambda z)} \, d\xi \, d\eta,$$

where

$$C(\mathbf{x}) = \frac{K}{z} e^{ikz} e^{ik(x^2+y^2)/2z},$$

and

$$g(\xi, \eta) = \psi_1(\xi, \eta) A(\xi, \eta) e^{ik(\xi^2 + \eta^2)/2z}.$$

We note that $\psi_1(\xi, \eta)$ is usually considered the optical disturbance just "behind" plane 1 and $\psi_1(\xi, \eta) A(\xi, \eta)$ the optical disturbance in, or just "ahead" of, plane 1. Equation (8.20), therefore, describes the propagation of $\psi_1(\xi, \eta) A(\xi, \eta)$ rather than that of $\psi_1(\xi, \eta)$. If we compare the integral in Equation (8.20) to that of Equation (3.26) (two-dimensional Fourier transform) we see that it is simply the Fourier transform of $g(\xi, \eta)$ with respect to the frequencies $(u, v) = (x/\lambda z, y/\lambda z)$. Thus we write Equation (8.19) as

$$(8.21) \qquad \psi_2(x, y) = C(x) \mathscr{F}_s \left[\psi_1(\xi, \eta) A(\xi, \eta) e^{(ik/2z)(\xi^2 + \eta^2)} \right].$$

$\mathscr{F}_s$ is called a scaled Fourier transform and this simply means that the transform frequencies must be scaled by $1/\lambda z$.

Let us review the operations dictated by Equation (8.21). *We begin with the optical disturbance function at plane* 1 $\psi_1(\xi, \eta) A(\xi, \eta)$:

1. *Multiply it by the quadratic phase term* $\exp[ik(\xi^2 + \eta^2)/2z]$ *to form the new function* $g(\xi, \eta)$,
2. *Compute the Fourier transform* $G(u, v)$ *of this new function,*
3. *Scale the transform frequencies by substituting* $x/\lambda z$ *and* $y/\lambda z$ *for* u *and* v, *respectively,*
4. *Multiply the scaled transform by the amplitude and phase terms*

$$C(x) = \frac{K}{z} e^{ikz} e^{ik(x^2 + y^2)/2z}$$

to obtain $\psi_2(x, y)$.

Equation (8.21) tells us that as the optical disturbance $\psi_1(\xi, \eta) A(\xi, \eta)$ propagates from plane 1 to plane 2 it diffracts, or changes, into a new form $\psi_2(x, y)$. For extremely large values of z, with respect to $(\xi^2 + \eta^2)$, the quadratic phase term $\exp[ik(\xi^2 + \eta^2)/2z]$ is approximately a constant and equal to unity. Thus we have

$$(8.22) \quad \psi_2(x, y) = \frac{K}{z} e^{ikz} e^{(ik/2z)(x^2 + y^2)} \mathscr{F}_s \left[\psi_1(\xi, \eta) A(\xi, \eta) \right] \quad \text{for large } z.$$

Therefore, we see as an optical disturbance propagates toward infinity it diffracts, or evolves, into its Fourier transform. This rather nebulous region of large z is known as the Fraunhofer, or far field, region and Equation (8.22) is called the Fraunhofer diffraction equation.

As our first example we study the propagation of a light beam that has a Gaussian spatial profile in plane 1. This is of interest for the following reasons. First, the output from most lasers is closely approximated by this shape. Second, this is one of the few problems that can be solved analytically. The Gaussian profile is described mathematically as

$$(8.23) \qquad \psi_1(\xi, \eta) = e^{-(\xi^2 + \eta^2)/\sigma^2},$$

where σ is known as the $1/e$ radius of the beam. That is, when $\xi^2 + \eta^2 = \sigma^2$ we have $\psi_1(\xi, \eta) = \exp[-1] = 1/e$. For this problem we choose a circular aperture function $A(\xi, \eta)$:

$$A(\xi, \eta) = \begin{cases} 1, & \xi^2 + \eta^2 \leqslant r^2, \\ 0 & \text{otherwise.} \end{cases}$$

Also let us choose the radius r of the aperture such that $r/\sigma = 4$. We note that $\psi_1(\xi, \eta)$ falls off rapidly with increasing distance $\xi^2 + \eta^2$ and, as a matter of fact, for $\xi^2 + \eta^2 \geqslant r^2$ we have

$$\psi_1(\xi, \eta) \leqslant e^{-4^2} = 1.13 \times 10^{-7}.$$

Thus in plane 1, we can say

$$\psi_1(\xi, \eta) A(\xi, \eta) \approx \psi_1(\xi, \eta).$$

Physically, this assumption means that the aperture in plane 1 does not significantly "clip" the beam. Mathematically, it means we now have a form that we can solve. We first use Equation (8.22) to determine the far field, or Fraunhofer, diffraction pattern

$$\psi_2(x, y) = \frac{K}{z} e^{ikz} e^{(ik/2z)(x^2 + y^2)} \mathcal{F}_s \left[e^{-(\xi^2 + \eta^2)/\sigma^2} \right].$$

Making use of the fact that this is a separable function (Chapter 3) and Equation (3.16), we can immediately write

$$(8.24) \qquad \psi_2(x, y) = \frac{\sigma^2 \pi}{z} K e^{ikz} e^{(ik/2z)(x^2 + y^2)} e^{-\pi^2 \sigma^2 (x^2 + y^2)/\lambda^2 z^2}.$$

Thus in the far field, the spatial profile is also Gaussian with $1/e$ radius $\lambda z/\pi\sigma$ and quadratic phase curvature $k(x^2 + y^2)/2z$.

We now calculate the shape of the beam for smaller values of z or what is commonly called the near field. To do so we use Equation (8.21).

$$\psi_2(x, y) = C(\mathbf{x}) \mathcal{F}_s \left[e^{-(\xi^2 + \eta^2)/\sigma^2} e^{ik((\xi^2 + \eta^2)/2z)} \right],$$

or

$$\psi_2(x, y) = C(\mathbf{x})\mathcal{F}_s\left[e^{-S(\xi^2+\eta^2)}\right],$$

where

$$s = \frac{1}{\sigma^2} - \frac{ik}{2z}.$$

Thus using the same techniques as before we obtain

$$\psi_2(x, y) = \frac{\pi}{s}C(\mathbf{x})e^{-\pi^2(\xi^2+\eta^2)/(\lambda^2 z^2 s)}.$$

Now substituting the proper expression for s in the preceding equation we obtain

(8.25) $$\psi_2(x, y) = B(\mathbf{x})e^{i\beta(x^2+y^2)}e^{-(x^2+y^2)/\alpha^2},$$

where the amplitude and linear phase term is

$$B(\mathbf{x}) = \frac{K}{z}\frac{e^{i(kz+\varphi)}}{(4z^2 + k^2\sigma^4)},$$

$$\tan\varphi = \frac{k\sigma^2}{2z}.$$

The $1/e$ radius is

$$\alpha = \frac{\lambda}{2\pi\sigma}\sqrt{4z^2 + k^2\sigma^4},$$

and the quadratic phase curvature is

$$\beta = \frac{2\pi^2 k\sigma^4}{\lambda^2 z(4z^2 + k^2\sigma^4)} + \frac{k}{2z}.$$

Thus we see at any location z (in the near field), the beam has a Gaussian profile and from this we conclude that a Gaussian beam will propagate through space while retaining its original shape. Note, however, that the dimensions and phase of the beam will change.

Next let us consider the propagation of a plane wave that passes through a rectangular aperture at plane 1. Mathematically, this is described as

$$\psi_1(\xi, \eta) = 1 \quad \text{and} \quad A(\xi, \eta) = \text{rect}_{ab}(\xi, \eta).$$

We first use Equation (8.22) to obtain the far field diffraction pattern

$$\psi_2(x, y) = \frac{K}{z} e^{ikz} e^{(ik/2z)(x^2+y^2)} \mathscr{F}_s \left[\text{rect}_{ab}(\xi, \eta) \right].$$

Using Equation (3.32) we are able to write

$$(8.26) \qquad \psi_2(x, y) = \frac{4ab}{z} e^{ikz} e^{(ik/2z)(x^2+y^2)} \text{sinc} \frac{2\pi ax}{\lambda z} \text{sinc} \frac{2\pi by}{\lambda z}.$$

We see, therefore, that as a rectangular profile propagates, it diffracts into a two-dimensional (separable) sinc function.

To determine the near field diffraction pattern we use Equation (8.21):

$$\psi_2(x, y) = C(\mathbf{x}) \mathscr{F}_s \left[e^{(ik/2z)(\xi^2+\eta^2)} \text{rect}_{ab}(\xi, \eta) \right].$$

Although this equation can be approached analytically by writing it in its integral form [Equation (3.26)] and using Fresnel integrals, we choose to calculate it on a computer using techniques discussed in Chapter 5. The normalized results of this computation for the case in which $a = b$ are shown (one-dimensionally) in Figures 8.8 through 8.10 in which the evolution from a rectangular profile to a two-dimensional sinc profile is readily apparent.

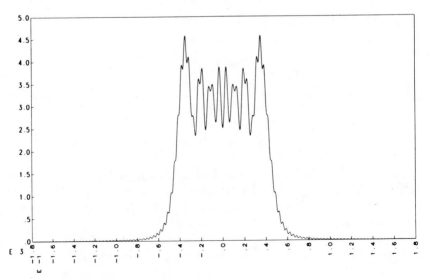

FIGURE 8.8. Near field diffraction from square aperture ($a/z = 1/50$).

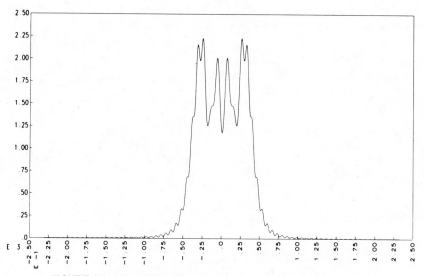

FIGURE 8.9. Near field diffraction from square aperture ($a/z = 1/100$).

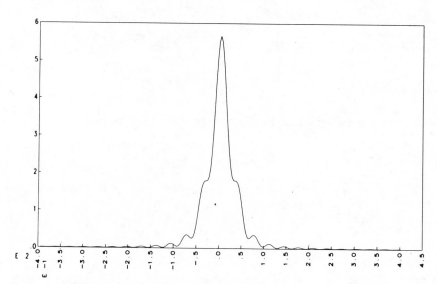

FIGURE 8.10. Near field diffraction from square aperture ($a/z = 1/100$).

211

We now derive the second Fourier transform form of the Fresnel equation. To do this we first rewrite Equation (8.18) as

$$\psi_2(x, y) = \frac{K}{z} e^{ikz} \iint_R \psi_1(\xi, \eta) g(x - \xi, y - \eta) \, d\xi \, d\eta,$$

where

$$g(\xi, \eta) = e^{ik(\xi^2 + \eta^2)/2z}.$$

We should recognize this equation as the two-dimensional convolution of $\psi_1(\xi, \eta)$ with $g(\xi, \eta)$:

$$\psi_2(x, y) = \frac{K}{z} e^{ikz} [\psi_1(x, y) * g(x, y)].$$

Using the two-dimensional convolution Theorem 3.21 we know that we can write this equation in the frequency domain as

(8.27) $$\Psi_2(u, v) = \frac{K}{z} e^{ikz} \Psi_1(u, v) G(u, v),$$

where $\Psi_1(u, v)$ and $\Psi_2(u, v)$ are the Fourier transforms of $\psi_1(x, y)$ and $\psi_2(x, y)$, respectively. Also, $G(u, v)$ is the Fourier transform of $g(x, y)$ which can be shown to be

(8.28) $$G(u, v) = \frac{-\lambda z}{i} e^{-i\pi\lambda z(u^2 + v^2)}.$$

Using this we can write Equation (8.27) as (recall that $K = -i/\lambda$)

(8.29) $$\Psi_2(u, v) = e^{ikz} \Psi_1(u, v) e^{-i\pi\lambda z(u^2 + v^2)}.$$

Equation (8.29) tells us that we can consider the propagation problem as a system with input $\psi_1(\xi, \eta)$ and output $\psi_2(x, y)$. The transfer function for this system is simply

(8.30) $$H(u, v) = e^{ikz} e^{-i\pi\lambda z(u^2 + v^2)}.$$

Thus to calculate the optical disturbance $\psi_2(x, y)$ at plane 2 resulting from the optical disturbance $\psi_1(\xi, \eta)$ at plane 1 we proceed as follows:

1. *Compute $\Psi_1(u, v)$, the Fourier transform of $\psi_1(\xi, \eta)$.*
2. *Multiply $\Psi_1(u, v)$ by the free space transfer function $H(u, v)$ of Equation (8.30) to obtain $\Psi_2(u, v)$.*
3. *Obtain $\psi_2(x, y)$ by inverse Fourier transforming $\Psi_2(u, v)$.*

To illustrate this technique we study the diffraction from two point sources separated by a distance $2a$. We choose our coordinate system so that the point sources are located in plane 1 as shown in Figure 8.11. Mathematically, a point source of light is described by a two-dimensional separable delta function. Notationally, we write

$$\delta(x, y) = \delta(x)\delta(y).$$

Thus our two point sources in plane 1 are described as

$$\psi_1(\xi, \eta) = \delta(\xi - a, \eta) + \delta(\xi + a, \eta).$$

The Fourier transform of this function is (see Chapter 3)

$$\Psi_1(u, v) = e^{2\pi iua} + e^{-2\pi iua}.$$

Using Equation (8.29) we have

(8.31) $\qquad \Psi_2(u, v) = e^{ikz}\left(e^{2\pi iua} + e^{-2\pi iua}\right)e^{-i\pi\lambda zu^2}e^{-i\pi\lambda zv^2}.$

Making use of the first shifting theorem 3.2 in the x (or u) dimension we find

(8.32) $\quad \psi_2(x, y) = \dfrac{K}{z}e^{ikz}\left[e^{(ik/2z)(x-a)^2} + e^{(ik/2z)(x+a)^2}\right]e^{(ik/2z)y^2}.$

However,

$$e^{(ik/2z)(x-a)^2} + e^{(ik/2z)(x+a)^2} = 2e^{(ik/2z)(x^2+a^2)}\cos\frac{kax}{z},$$

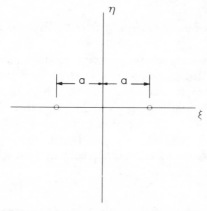

FIGURE 8.11. Point sources separated by a distance $2a$.

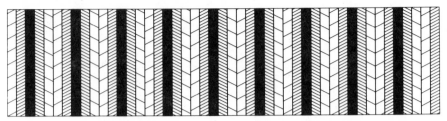

FIGURE 8.12. Diffraction pattern from two point sources.

and thus our equation becomes

$$(8.33) \qquad \psi_2(x, y) = 2K\frac{e^{ikz}}{z}e^{(ik/2z)(x^2+y^2+a^2)}\cos\frac{2\pi ax}{\lambda z}.$$

The modulus squared of the preceding equation is given as

$$\|\psi_2(x, y)\|^2 = \frac{4K^2}{z^2}\cos^2\frac{2\pi ax}{\lambda z}$$

and thus in plane 2 we see "vertical" interference fringes as illustrated in Figure 8.12. The spacing between any two fringes is $\lambda z/2a$.

In this section we have shown how both forms of the Fresnel diffraction equation can be "solved" using Fourier transforms. More often than not, these solutions will have to be obtained using the techniques presented in Chapter 5. Therefore, we now comment upon the relative advantages (and disadvantages) of each form.

Equation (8.21). Fourier Form of Second Fresnel Equation (8.19)

This method is attractive because it requires only one Fourier transform operation. The main disadvantage of this method lies mostly in the fact that the function to be transformed often has very high frequency content and, therefore, a digital solution requires a very large number of samples. As can be seen from Equation (8.21), the function to be transformed is the product of $\psi_1(\xi,\eta)$, $A(\xi, \eta)$, and $\exp[ik(\xi^2 + \eta^2)/2z]$ and, because k (in the optical spectrum) is a very large number, the frequency content is normally dominated by the quadratic phase term

$$e^{ik(\xi^2+\eta^2)/2z} = e^{2\pi i\xi^2/2\lambda z}e^{2\pi i\eta^2/2\lambda z}.$$

For the sake of clarity, we consider each dimension individually. First, the ξ-dimension,

$$e^{2\pi i \xi^2/2\lambda z} = e^{2\pi i w_\xi \xi},$$

where we have expressed the term $\xi/2\lambda z$ as a linearly varying frequency term w_ξ. The maximum value of this frequency obviously occurs at the maximum value of ξ (ξ_{max}) and is given by

$$w_{max} = \frac{\xi_{max}}{2\lambda z}.$$

We learned in Chapter 5 (real world sampling theorem 5.2) that to properly digitize a function we should sample it about 10 times per cycle of the highest frequency present. In our case this becomes $\Delta\xi = 2\lambda z/10\xi_{max}$. Therefore, if we sample the function from $-\xi_{max}$ to ξ_{max} we obtain N_ξ samples, where

$$N_\xi = \frac{2\xi_{max}}{\Delta\xi} = \frac{10\xi_{max}^2}{\lambda z}.$$

Similarly, in the η-dimension we require N_η samples, where

$$N_\eta = \frac{2\eta_{max}}{\Delta\eta} = \frac{10\eta_{max}^2}{\lambda z}.$$

For example, let us suppose $\xi_{max} = \eta_{max} = 0.1$ cm, $z = 10$ cm, and $\lambda = 0.6328 \times 10^{-4}$ cm, then $N_\xi = N_\eta = 158$. Thus to completely sample the function in two dimensions we require $N_\xi N_\eta = (158)(158) = 24,964$ samples.

Equation (8.29). Fourier Form of First Fresnel Equation (8.18)

The main disadvantage of this method is that it requires two Fourier transform operations. However, the number of samples required using this method is often significantly less than that of the previously discussed method. The following discussion should clarify these remarks.

To apply Equation (8.29) we must first transform $\psi_1(\xi, \eta)$ which, in the ξ-dimension, requires a sampling rate $\Delta\xi = 1/(10u_{max})$ where u_{max} is obviously the maximum ξ-dimension frequency content. Therefore, sampling the function from $-\xi_{max}$ to ξ_{max} requires $20\xi_{max}u_{max}$ samples. Using analogous reasoning in the η-dimension and multiplying the results we determine the total number of samples required (two-dimensional) to be

$$(8.34) \qquad N_1 = 400\xi_{max}\eta_{max}u_{max}v_{max}.$$

The next step is to multiply this transform by the free space transfer function $H(u, v)$ [Equation (8.30)] and then inverse transform the resulting product. $H(u, v)$ in the u-dimension has a maximum frequency of $\lambda z u_{max}/2$ and, therefore, requires $10\lambda z u_{max}$ samples. Similar remarks hold true for the v-dimension. Thus the total number of samples required for the second transform operation is the greater of N_1 or

$$(8.35) \qquad N_2 = 100(\lambda z u_{max} v_{max})^2.$$

Note that inasmuch as λ^2 appears in this expression, it tends to reduce the value of N_2.

As a comparative example, let us determine the number of samples required to digitally propagate the Gaussian beam

$$\psi(\xi, \eta) = e^{-(\xi^2 + \eta^2)},$$

over the aperture

$$A(\xi, \eta) = \{(\xi, \eta)| - 2.7 \leqslant \xi \leqslant 2.7, -2.7 \leqslant \eta \leqslant 2.7\},$$

a distance 20 cm to plane 2. The wavelength is 0.6328×10^{-4} cm. The Fourier transform of $\psi(\xi, \eta)A(\xi, \eta)$ is the "almost bandlimited" function

$$\psi(u, v) = \pi e^{-\pi^2(u^2 + v^2)}$$

with bandlimits $u_{max} = v_{max} = 1.05$ cm^{-1}. For method 2 [Equation (8.29)] we calculate (Equations 8.34–8.35)

$$N_1 = 400(2.7)(2.7)(1.05)(1.05) = 3215,$$

$$N_2 = 100[(0.6328 \times 10^{-4})(20)(1.05)(1.05)]^2 < 1.$$

Thus to use method 2 we require 3215 for the first transform operation and also 3215 for the second inverse transform operation. Using the first method [Equation (8.21)] we determine

$$N_\xi = N_\eta = \frac{(10)(2.71)^2}{(20)(0.6328 \times 10^{-4})} = 57601.$$

Thus to do the job in two dimensions requires $N_\xi N_\eta > 3 \times 10^9$ samples. Now if the propagation distance is increased from 20 to 20,000 cm, we find that method 2 requires $N_1 = 3215$ and $N_2 = 195$. Therefore, we still require 3215 samples for each transform operation. On the other hand, for method 1 we have $N_\xi N_\eta = 3318$. We see that the number of samples are approximately the same for this case and, therefore, method 1 should be used inasmuch as it only requires one transform operation.

Note that $N_\xi N_\eta$ for method 1 is the number of samples required to properly digitize the quadratic phase term, which is usually the dominating factor. However, it is possible that the number of samples required to digitize $\psi(\xi, \eta)A(\xi, \eta)$ [also equal to N_1 in Equation (8.29)] will be larger. A little reflection upon this fact reveals that method 1 will never require fewer samples than method 2, but all things being equal it is the preferred method because it only requires one transform operation.

THIN LENSES AND DIFFRACTION

Up to this point in the chapter we have shown how to mathematically describe the propagation of light from one plane to another. We learned that an optical disturbance $\psi(\xi, \eta)$ in plane 1 will diffract, or evolve, into its Fourier transform (far field diffraction pattern) as it propagates toward infinity. In this section we show that the effect of placing a lens in (or just ahead of) plane 1 is to either speed up or slow down this diffraction process.

We define a positive thin lens as an optical element that modifies the phase of an optical disturbance. Such a lens is shown schematically in Figure 8.13 in which the input optical disturbance is denoted as $\psi_i(x, y)$, and the output as $\psi_o(x, y)$. The relationship between them is

$$(8.36) \quad \psi_o(x, y) = \psi_i(x, y)Le^{(-ik/2f)(x^2+y^2)}, \quad \text{where } L = e^{2\pi in\Delta_o/\lambda}.$$

In this equation n is the index of refraction of the lens material, Δ_o is the on-axis thickness, and f is the focal length of the lens. *We will always consider f to be a positive number* (later when we discuss negative lenses we use $-f$).

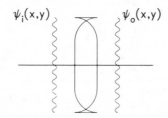

FIGURE 8.13. Thin lens of a phase modulator.

From Equation (8.36) we can see that the effect of a lens is to add quadratic phase curvature to the incident optical wave. Also in this equation we note the fact that the lens has no effect on the amplitude of $\psi(x, y)$. This is known as the infinite lens assumption and simply means that the aperture (or diameter) of the lens is larger than that of the incident light beam. When this assumption is not valid and we wish to take into account the finite diameter of the lens, we simply add an aperture term to the preceding equation. In much of our work we consider the lens to be infinite.

We now proceed to determine what effect a lens placed in plane 1 will have on the optical disturbance function as it propagates toward infinity. The geometry of this situation is shown in Figure 8.14. Just before the lens is the optical disturbance $\psi_1(\xi, \eta)A(\xi, \eta)$ which we are interested in propagating through the lens and "downstream" a distance z. Using Equation (8.36) we determine the optical disturbance function in plane 1 just after the lens to be

$$\psi_1(\xi, \eta)A(\xi, \eta)Le^{(-ik/2f)(\xi^2+\eta^2)}$$

Now using Equation (8.21) to propagate this disturbance to plane 2 we obtain

$$(8.37) \quad \psi_2(x, y) = LC(\mathbf{x})\mathscr{F}_s\left[\psi_1(\xi, \eta)A(\xi, \eta)e^{(ik/2)(1/z-1/f)(\xi^2+\eta^2)}\right]$$

Now if we let

$$(8.38) \qquad\qquad \hat{z} = \frac{zf}{f - z}$$

in the preceding equation we obtain

$$(8.39) \qquad \psi_2(x, y) = LC(\mathbf{x})\mathscr{F}_s\left[\psi_1(\xi, \eta)A(\xi, \eta)e^{ik(\xi^2+\eta^2)/2\hat{z}}\right].$$

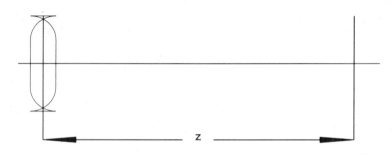

Plane 1   Plane 2

FIGURE 8.14. Thin lens placed in plane 1.

Equation (8.39) looks almost like the propagation of $\psi_1(\xi, \eta)A(\xi, \eta)$ a distance $\hat{z}$ without a lens. We say *almost* because the preceding equation uses the scale factor $1/\lambda z$ rather than $1/\lambda \hat{z}$ and also the phase factor $C(\mathbf{x})$ contains z rather than $\hat{z}$. This situation is graphically illustrated in Figure 8.15 where (a) shows the propagation of the disturbance $\psi_1(\xi, \eta)A(\xi, \eta)$ through the lens and then to plane 2 a distance z away. In (b) we show the propagation of this same disturbance through free space a distance $\hat{z}$ to plane 3. Using Equations (8.21) and (8.39) we are able to determine

$$(8.40) \qquad \psi_2(x, y) = sN(x, y, z, \hat{z})\psi_3(sx, sy),$$

where

$$N(x, y, z, \hat{z}) = Le^{ik(z-\hat{z})}e^{(ik/2)(x^2+y^2)(z-\hat{z})/(z\hat{z})} \quad \text{and} \quad s = \hat{z}/z.$$

From this equation we see that $\psi_2(x, y)$ and $\psi_3(x, y)$ are isomorphic in the sense that

$$\|\psi_2(x, y)\| = \left|\frac{\hat{z}}{z}\right| \|\psi_3(sx, sy)\|.$$

Thus for every distance z with a lens (Figure 8.15a), there is a distance $\hat{z}$ without a lens (Figure 8.15b) such that the two resulting diffraction patterns

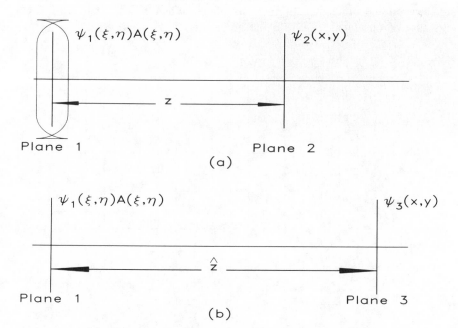

FIGURE 8.15. Propagation isomorphism with and without lens.

have the same basic profile. The relationship between these two distances is given by the simple algebraic equation (8.38). To determine the diffraction pattern $\psi_2(x, y)$ at some distance z past a lens we (1) determine $\hat{z}$ [Equation (8.38)], (2) calculate the free space diffraction pattern $\psi_3(x, y)$ at $\hat{z}$ [Equation (8.21) or (8.29)], and then (3) use Equation (8.40) to relate the two.

We now proceed to use Equation (8.38) by itself to develop some other interesting and well-known results. We begin by normalizing the variables in Equation (8.38) by the lens focal length f:

$$(8.41) \qquad \frac{\hat{z}}{f} = \frac{z/f}{1 - z/f} \quad \text{(positive lens)}$$

which is plotted in Figure 8.16. In Figure 8.17 we have graphically illustrated the isomorphism of Equation (8.41) for $0 \leqslant z/f \leqslant 1$. As can be seen, when $z/f = 1$ we have $\hat{z}/f = \infty$ or, in other words, at one focal length past the lens the diffraction pattern is isomorphic to the far field, or Fraunhofer, diffraction pattern. Inasmuch as the optical disturbance at the lens evolves into its Fourier transform in a distance of only one focal length, we say that a positive lens "speeds up" the diffraction process. Later in the chapter we discuss this Fourier transforming property of a positive lens in greater detail. In Figure 8.18 we present another graphical illustration of Equation (8.41) isomorphism, only this time for $z/f > 1$. As can be appreciated from this figure, when the propagation distance z becomes greater than the lens focal length, the sign of $\hat{z}$ changes and the isomorphic propagation distance starts backwards from infinity on the other side of the lens plane. When $z = \infty$, $\hat{z}$ is equal to a distance of one focal length on the other side of the lens. When $z = 2f$, we

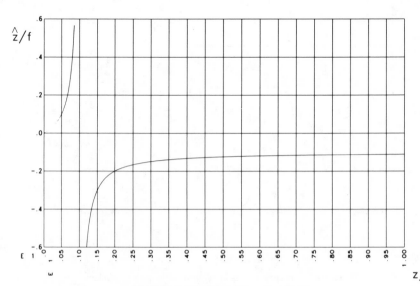

FIGURE 8.16.   Plot of Equation (8.41).

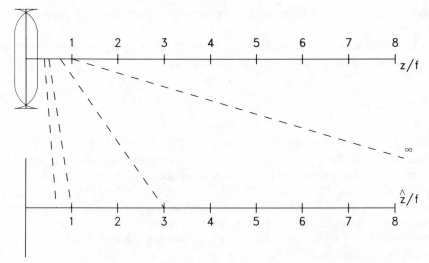

FIGURE 8.17. Graphical representation of Equation (8.41) for $0 \leqslant z/f \leqslant 1$.

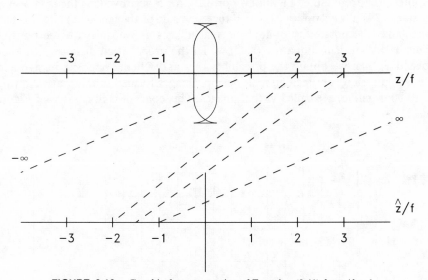

FIGURE 8.18. Graphical representation of Equation (8.41) for $z/f > 1$.

have $\hat{z} = -2f$ and thus we obtain an imaging condition. In general, we have imaging whenever the isomorphic diffraction pattern $\psi_3(x, y)$ lies on the opposite side of the lens from $\psi_2(x, y)$. Thus, mathematically, for imaging we require $\hat{z} = -s$. Now using this fact in Equation (8.38) we obtain the simple lens equation of geometrical optics:

$$-s = \frac{zf}{f-z} \quad \text{or} \quad \frac{1}{s} + \frac{1}{z} = \frac{1}{f}.$$

In the next section we have much more to say about the imaging properties of a lens.

For a negative or concave lens we simply replace f by $-f$ in all our previous equations. Doing so in Equation (8.41) we obtain

$$(8.42) \qquad \frac{\hat{z}}{f} = \frac{z/f}{1 + z/f} \quad \text{(negative lens)}.$$

This equation is plotted in Figure 8.19 and its isomorphism graphically illustrated in Figure 8.20 for $z/f \geqslant 0$. As can be seen in these figures, the isomorphic propagation distance $\hat{z}$ can never become greater than the focal length of the lens. That is, the diffraction pattern $\psi_2(x, y)$ of the system with a negative lens can never develop further than the isomorphic diffraction pattern propagating a distance f with no lens. Such a system is very useful in the design of coherent systems in which it is desired to minimize the diffraction "breakup" of a propagating beam.

As our last topic in this section we discuss virtual images. As can be appreciated from either Figure 8.19 or 8.20, for a negative lens the isomorphic distance $\hat{z}$ is always positive (i.e., on the same side of the lens as z) and this fact precludes the possibility of forming real images as we did with the positive lens. However, if in Equation (8.42) we let the propagation distance z become negative, then we obtain the possibility of z and $\hat{z}$ being opposite in sign and thus establishing an imaging relationship. Images formed in this way (allowing z to be negative) are called virtual images. This concept is illustrated in Figure 8.21.

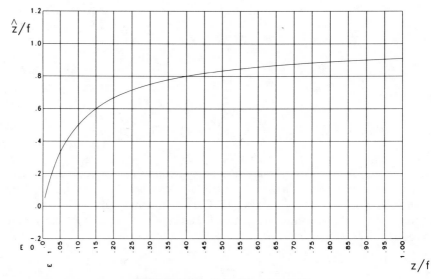

FIGURE 8.19. Plot of Equation (8.42).

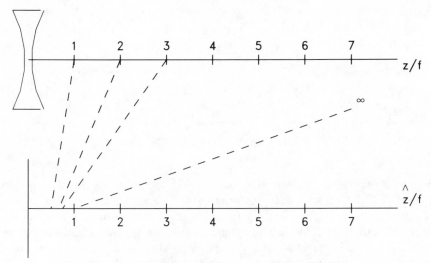

FIGURE 8.20.   Graphical representation of Equation (8.42) for $z/f \geqslant 0$.

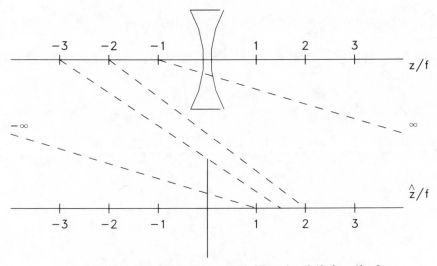

FIGURE 8.21.   Graphical representation of Equation (8.42) for $z/f < 0$.

IMAGING PROPERTIES OF LENSES

In the previous section we discussed the effect of a lens on diffraction. In the course of this study we introduced the concept of an isomorphic propagation distance from which we were able to obtain the simple lens equation of geometrical optics. In this section we again derive this equation. However, this time we use an approach that provides much more information than just this

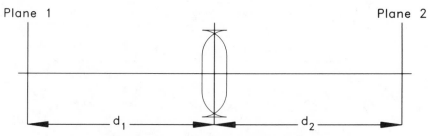

FIGURE 8.22. Imaging with a single positive lens.

imaging relationship. Shown in Figure 8.22 is the geometry we discuss. An optical disturbance in plane 1 propagates a distance d_1 through free space, passes through a positive lens (of focal length f), and then continues on the distance d_2 to plane 2. We let the optical disturbance functions be described by $\psi_1(x_1, y_1)$ in plane 1 and $\psi_2(x_2, y_2)$ in plane 2. Also, we consider it to be given by $\psi_-(x, y)$ just before and $\psi(x, y)$ just after the lens. We first use Equation (8.21) to propagate the beam from plane 1 up to the lens:

$$\psi_-(x, y) = \frac{K}{d_1} e^{ikd_1} e^{ik(x^2+y^2)/2d_1} \mathcal{F}_{s1}\left[\psi_1(x_1, y_1) e^{(ik/2d_1)(x_1^2+y_1^2)}\right].$$

Now using Equation (8.36) we determine the disturbance $\psi(x, y)$ just after the lens to be

$$\psi(x, y) = \frac{LK}{d_1} e^{ikd_1} e^{(ik/2)[(1/d_1)-(1/f)](x^2+y^2)} \mathcal{F}_{s1}\left[\psi_1(x_1, y_1) e^{(ik/2d_1)(x_1^2+y_1^2)}\right].$$

Finally, we propagate this disturbance to plane 2 again using Equation (8.21):

(8.43)

$$\psi_2(x_2, y_2) = \frac{K^2 L}{d_1 d_2} e^{ik(d_1+d_2)} e^{(ik/2d_2)(x_2^2+y_2^2)}$$

$$\times \mathcal{F}_{s2}\left[e^{(ik/2)[(1/d_1)+(1/d_2)-(1/f)](x^2+y^2)} \mathcal{F}_{s1}\left\{\psi_1(x_1, y_1) e^{(ik/2d_1)(x_1^2+y_1^2)}\right\}\right]$$

The preceding equation describes the optical disturbance $\psi_2(x_2, y_2)$ resulting from the propagation of the disturbance $\psi_1(x_1, y_1)$ from plane 1 through the lens to plane 2. Thus far no restrictions have been placed on the distances d_1 and d_2. Now, however, let us restrict them so they satisfy the geometrical imaging relationship

$$\frac{1}{d_1} + \frac{1}{d_2} = \frac{1}{f}.$$

When this assumption is used in Equation (8.43) we obtain

$$(8.44) \quad \psi_2(x_2, y_2) = \hat{C}(\mathbf{x}_2)\mathscr{F}_{s2}\Big\{\mathscr{F}_{s1}\Big[\psi_1(x_1, y_1)e^{(ik/2d_1)(x_1^2+y_1^2)}\Big]\Big\},$$

where

$$\hat{C}(\mathbf{x}_2) = \frac{K^2L}{d_1d_2}e^{ik(d_1+d_2)}e^{ik(x_2^2+y_2^2)/2d_2}.$$

The problem now becomes that of determining the scaled Fourier transform of a scaled Fourier transform. For the sake of clarity, we show this result in one dimension for the function $f(x)$ which has a Fourier transform denoted as $F(w)$. We first note

$$\mathscr{F}_{s1}[f(x)] = F\Big(\frac{w}{\lambda d_1}\Big).$$

Next we use Theorem 3.6 (transform of a transform) and the scaling Theorem 3.5 (with $a = 1/\lambda d_1$) to determine

$$\mathscr{F}F\Big(\frac{w}{\lambda d_1}\Big) = \lambda d_1 f(-x\lambda d_1),$$

$$\mathscr{F}_{s2}F\Big(\frac{w}{\lambda d_1}\Big) = \lambda d_1 f\Big(\frac{-xd_1}{d_2}\Big).$$

Now generalizing this result to two dimensions and using it in Equation (8.44) we obtain (*Note*: $x_2 = y_1d_1/d_2$ and $y_2 = x_1d_1/d_2$)

$$(8.45)$$

$$\psi_2(x_2, y_2) = -\frac{d_1}{d_2}Le^{ik(d_1+d_2)}e^{ik(d_1+d_2)(x_2^2+y_2^2)/(2d_1d_2)}\psi_1\Big(\frac{-x_1d_1}{d_2}, \frac{-y_1d_1}{d_2}\Big).$$

If we let $M = d_2/d_1$ and consider just the modulus squared (what we physically see) of both sides of the preceding equation we find

$$(8.46) \quad \|\psi_2(x_2, y_2)\|^2 = \frac{1}{M^2}\Big\|\psi_1\Big(\frac{-x_1}{M}, \frac{-y_1}{M}\Big)\Big\|^2.$$

Therefore, we see that $\psi_2(x_2, y_2)$ is simply an image of the disturbance in plane 1 with magnification M. Note the negative sign on x_1 and y_1 that corresponds to the fact that images formed in this way (with a single lens) are inverted. In addition to image information such as magnification and amplitude, we can also use Equation (8.45) to learn about the phase of the "image disturbance" at plane 2.

We consider the image to be out of focus when the imaging relationship is not satisfied. That is to say, when

$$\frac{1}{d_1} + \frac{1}{d_2} - \frac{1}{f} = \varepsilon \neq 0.$$

When this is the case, geometrical optics can only tell us that we do not have a (perfect) image. However, using Equation (8.43)—and performing the transforms digitally—we can still predict what this defocused image will look like.

In the previous development we have used the infinite lens assumption. That is, we have assumed that the diameter of the lens was infinite and, therefore, did not "clip," or aperture, the light as it propagated through. Obviously, real lenses are not available this large and, therefore, we must modify our analysis. Let us assume that the lens does, in fact, have a finite aperture described by $A_L(x, y)$. Therefore, the lens modulation equation (8.36) is modified to read

$$\psi_o(x, y) = \psi_i(x, y) L A_L(x, y) e^{-(ik/2f)(x^2+y^2)}.$$

When this is taken into account in our previous development, Equation (8.43) becomes

$$(8.47) \quad \psi_2(x_2, y_2) = \frac{K^2 L}{d_1 d_2} e^{ik(d_1+d_2)} e^{(ik/2d_2)(x_2^2+y_2^2)}$$

$$\times \mathcal{F}_{s2}\bigg\{ A_L(x, y) e^{(ik/2)(1/d_1 + 1/d_2 - 1/f)(x^2+y^2)}$$

$$\times \mathcal{F}_{s1}\bigg[\psi_1(x_1, y_1) e^{(ik/2d_1)(x_1^2+y_1^2)} \bigg] \bigg\}$$

Also, when the imaging relationship is satisfied, Equation (8.44) is modified to read

$$(8.48) \quad \psi_2(x_2, y_2) = \hat{C}(\mathbf{x}_2) \mathcal{F}_{s2}\bigg\{ A_L(x, y) \mathcal{F}_{s1}\bigg[\psi_1(x_1, y_1) e^{(ik/2d_1)(x_1^2+y_1^2)} \bigg] \bigg\}$$

where

$$\hat{C}(\mathbf{x}_2) = \frac{K^2 L}{d_1 d_2} e^{ik(d_1+d_2)} e^{ik(x_2^2+y_2^2)/2d_2}.$$

If now in the preceding equation we let

$$g(x, y) = \mathcal{F}_{s1}\left[\psi_1(x_1, y_1)e^{(ik/2d_1)(x_1^2+y_1^2)}\right]$$

we have

$$\psi_2(x_2, y_2) = \hat{C}(\mathbf{x}_2)\mathcal{F}_{s2}[A_L(x, y)g(x, y)].$$

The two-dimensional product Theorem 3.22 can be applied to this equation to obtain

$$\psi_2(x_2, y_2) = \hat{C}(\mathbf{x}_2)\mathcal{F}_{s2}[A_L(x, y)] * \mathcal{F}_{s2}[g(x, y)].$$

However, we have previously shown that

$$\mathcal{F}_{s2}[g(x, y)] = \lambda^2 d_1^2 \psi_1\left(\frac{-x_1}{M}, \frac{-y_1}{M}\right)e^{(ik/2d_1)(x_1^2+y_1^2)/M^2}$$

and, therefore, we arrive at the expression

(8.49)

$$\psi_2(x_2, y_2) = \frac{-d_1}{d_2}Le^{ik(d_1+d_2)}e^{(ik/2d_2)(x_2^2+y_2^2)}$$

$$\times \left(\mathcal{F}_{s2}[A_L(x, y)] * \psi_1\left(\frac{-x_1}{M}, \frac{-y_1}{M}\right)e^{(ik/2d_1)(x_1^2+y_1^2)/M^2}\right).$$

From this equation we see that our image distribution $\psi_2(x_2, y_2)$ is simply the ideal image distribution $\psi_1(-x_1/M, -y_1/M)$ convolved (or "smeared") with the scaled Fourier transform of the lens aperture function $A_L(x, y)$.

Much can be learned about optical systems in general by an examination of how they image a point source of light. In particular, let us examine how well our single (finite) lens will image a point source of light located at the origin of plane 1:

$$\psi_1(x_1, y_1) = \delta(x_1, y_1).$$

Using this, and recalling the fact that the delta function is the unit element under convolution, Equation (8.49) yields

(8.50) $$\psi_2(x_2, y_2) = \frac{-d_1}{d_2}Le^{ik(d_1+d_2)}e^{(ik/2d_2)(x_2^2+y_2^2)}\mathcal{F}_{s2}[A_L(x, y)],$$

or

(8.51) $$\|\psi_2(x_2, y_2)\|^2 = \frac{1}{M^2}\|\mathcal{F}_{s2}[A_L(x, y)]\|^2.$$

Thus we see that the image of this point is determined by the scaled Fourier transform of the lens aperture function. For example, let us examine how a circular lens of diameter a will image a point source of light. Also, for the sake of argument, let us assume that the lens is set up to image with equal conjugate distances (i.e., $d_1 = d_2 = 2f$). In this situation it behooves us to take advantage of the circular, or radial, symmetry of the system and use polar coordinates to describe our lens aperture function:

$$A_L(x, y) = \begin{cases} 1, & x^2 + y^2 \leqslant a \\ 0, & \text{otherwise} \end{cases}$$

or

$$A_L(r) = \begin{cases} 1, & r \leqslant a \\ 0, & \text{otherwise.} \end{cases}$$

We learned in Chapter 3 that the Fourier (Hankel) transform of this function is also circularly symmetric and given by

$$\mathscr{F}[A_L(r)] = \frac{aJ_1(2\pi wa)}{w},$$

where J_1 is the first-order Bessel function. Thus

$$\mathscr{F}_{s2}[A_L(r)] = \frac{aJ_1(2\pi r_2 a/\lambda d_2)}{r_2/\lambda d_2}$$

and Equation (8.51) yields

$$\|\psi_2(r_2)\|^2 = \left\| \frac{aJ_1(2\pi r_2 a/\lambda d_2)}{r_2/\lambda d_2} \right\|^2.$$

The preceding equation describes an Airy pattern and its shape is shown (normalized form) in Figure 8.23. We see from this figure that a finite circular lens will smear the image of a perfect point into a set of concentric rings. As a measure of this smear we often use the radius of the central ring; that is, the radius out to the first zero of the function. It can be seen from this figure that this value occurs at (about) 3.8 and thus

$$\hat{r}_2 \approx \frac{3.8\lambda d_2}{2\pi a}.$$

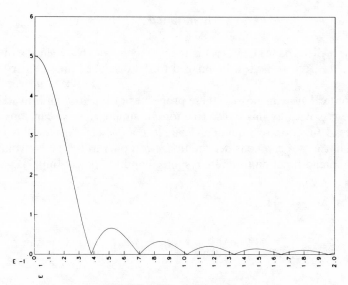

FIGURE 8.23. An Airy pattern.

A typical laboratory situation might consist of $a = 5$ cm, $d_2 = 20$ cm, $\lambda = 0.6328 \times 10^{-4}$ cm. Thus $\hat{r}_2 \approx 1.53 \times 10^{-4}$ cm.

Let us briefly summarize our previous development. We first assumed that the disturbance function at plane 1 was a point source or two-dimensional delta function. We then determined the image of this point in plane 2 which turned out to be an Airy pattern. This approach suggests that we treat our lens as a system with input function $\psi_1(x_1, y_1)$ and resulting output $\psi_2(x_2, y_2)$. When we do this (based on remarks in Chapter 6), we see that Equation (8.50) is the impulse response of the system. In optics, the impulse response is also called the point spread function. We learned in Chapter 6 that under proper conditions the impulse response of a system completely describes the system. These proper conditions are linearity and invariance. Optically, these properties are described as follows.

Linearity

If the objects $\psi_1(x_1, y_1)$ and $\varphi_1(x_1, y_1)$ are imaged to $\psi_2(x_2, y_2)$ and $\varphi_2(x_2, y_2)$, respectively, then any linear combination of these objects $a\psi_1(x_1, y_1) + b\varphi_1(x_1, y_1)$ will be imaged to the same linear combination of the individual images $a\psi_2(x_2, y_2) + b\varphi_2(x_2, y_2)$.

Invariance

If the object disturbance $\psi(x_1, y_1)$ is imaged to $\psi_2(x_2, y_2)$, then the displaced object $\psi(x_1 - \xi, y_1 - \eta)$ will be imaged to the displaced image $\psi_2(x_2 - \xi, y_2 - \eta)$.

Most optical systems possess these properties of linearity and invariance to a large degree. Because this is also true for our simple lens, we can consider the point spread function of Equation (8.50) to completely describe this imaging system. That is, we can consider the image (output) to be the convolution of the point spread function (impulse response) and the object (input):

$$(8.52) \quad \psi_2(x_2, y_2)$$

$$= -\frac{d_1}{d_2} L e^{ik(d_1 + d_2)} \left\{ e^{(ik/2d_2)(x_2^2 + y_2^2)} \mathcal{F}_{s2} [A_L(x, y)] \right\} * \psi_1(x_1, y_1).$$

It is instructive to compare this equation to Equation (8.49) which also obtains the image by a convolution operation on the object. In Equation (8.52) we have used to our advantage the fact that our system is linear and invariant and thus this equation appears slightly different than Equation (8.49). We note that the quadratic phase term $\exp[ik(x_1^2 + y_1^2)/(2M^2d_1)]$ of Equation (8.49) is not present in Equation (8.52). It would also be nice if we could dispense with the other quadratic term $\exp[ik(x_2^2 + y_2^2)/2d_2]$. It turns out that for most real world situations we can, indeed, remove this term. This can be illustrated by noting that, for most lenses, the term $\mathcal{F}_{s2}[A_L(x, y)]$ has a form similar to an Airy pattern (e.g., a two-dimensional sinc function for a rectangular lens aperture). For these patterns we define the characteristic dimension r_c to be the radius out to the first zero. This distance r_c has values that are of the order of $\lambda d_2/2a$. The value of the quadratic phase term at this characteristic dimension is

$$e^{(2\pi i/\lambda d_2)r_c^2} = e^{i\pi \lambda d_2/2a^2}.$$

If $\pi \lambda d_2/2a^2$ is sufficiently small, we may consider the phase term to be constant and thus the product term

$$e^{(ik/2d_2)(x_2^2 + y_2^2)} \mathcal{F}_{s2}[A_L(x, y)]$$

is dominated by, or essentially equal to $\mathcal{F}_{s2}[A_L(x, y)]$. Again, a typical situation might consist of $\lambda = 0.6328 \ 10^{-4}$ cm, $d_2 = 20$ cm, $a = 5$ cm, which yields $\pi \lambda d_2/2a^2 < 8 \times 10^{-5}$. When we drop the quadratic phase term, Equation (8.52) reduces to the simpler form

$$(8.53) \quad \psi_2(x_2, y_2) = -\frac{d_1}{d_2} e^{ik(d_1+d_2)} L \mathscr{F}_{s2} [A_L(x, y)] * \psi_1(x_1, y_1).$$

The preceding equation also provides us with a simpler form of the system's point spread function, namely,

$$(8.54) \quad h(x_2, y_2) = -\frac{d_1}{d_2} L e^{ik(d_1+d_2)} \mathscr{F}_{s2} [A_L(x, y)].$$

Obviously, we could have obtained this same result by removing the quadratic phase terms from Equation (8.50).

We know from Chapter 6 that the transfer function of our system is just the Fourier transform of its point spread function (impulse response function);

$$(8.55) \quad H(u, v) = \lambda^2 d_1 d_2 L e^{ik(d_2+d_1)} A_L(-\lambda d_2 u, -\lambda d_2 v).$$

Equation (8.55) is a refreshingly nice result. It simply states that the transfer function of a lens is the lens aperture function scaled by $-\lambda d_2$. Therefore, the determination of a lens transfer function is a trivial matter.

To illustrate this result we examine how well our lens images a series of line pairs which are shown (one-dimensionally) in Figure 8.24. The width of each line is $2a$ and the periodic spacing is $4a$. This object is described mathematically as the convolution of a comb and a pulse function:

$$(8.56) \quad \psi_1(x_1, y_1) = p_a(x_1) * \text{comb}_{4a}(x_1).$$

The Fourier transform of this function is

$$(8.57) \quad \Psi_1(u, v) = 2a \, \text{sinc} \, 2\pi u a \, \text{comb}_{1/4a}(u)$$

which is shown in Figure 8.25. We know (Chapter 5) that the product of any function $F(w)$ with a $\text{comb}_{\Delta w}(w)$ is a sequence whose terms are given by values of the function at discrete locations, that is, $\{F(k\Delta w)\}$. Using this fact we can

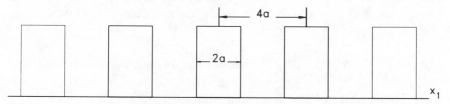

FIGURE 8.24. Line pairs to be imaged.

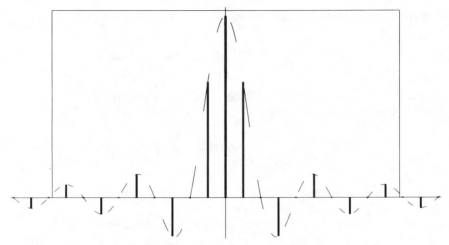

FIGURE 8.25.   Fourier transform of line pairs.

rewrite Equation (8.57) as

$$\Psi_1(u, v) = \left\{ 2a \operatorname{sinc} \frac{k\pi}{2} \right\}.$$

This distribution is, in fact, (4a times) the Fourier series representation of $\psi_1(x_1, y_1)$. This can easily be verified if we recall that the Fourier series coefficients of a periodic function can be obtained from the Fourier transform of that function (over one period) by using Equation (5.12):

$$C_k = \frac{1}{T} F\left(\frac{k}{T} \right).$$

For $p_a(x)$ (see Equation 3.11) we obtain

$$C_k = \frac{1}{2} \operatorname{sinc} \frac{k\pi}{2}.$$

The image Fourier transform $\Psi_2(u, v)$ is determined by the product of the lens transfer function $H(u, v)$ [Equation (8.55)] and the object Fourier transform (Equation 8.57):

$$\Psi_2(u, v) = 2a\lambda^2 d_1 d_2 L e^{ik(d_1 + d_2)} A_L(-\lambda d_2 u, -\lambda d_2 v) \operatorname{sinc} 2\pi u a \operatorname{comb}_{1/4a}(u).$$

If we assume that our lens is circular with radius r_c, then the lens aperture function is described by a circ function. However, because of the one-dimensional nature of our problem, we need only consider the x (or u) dimension and thus we obtain a simple pulse function. The equation then becomes

$$(8.58) \quad \Psi_2(u, v) = 2a\lambda^2 d_1 d_2 L e^{ik(d_1 + d_2)} p_{r_c}(-\lambda d_2 u)\text{sinc}(2\pi ua)\text{comb}_{1/4a}(u).$$

To obtain the image function $\psi_2(x_2, y_2)$ we could calculate the inverse Fourier transform of Equation (8.58). However, based on our previous comments concerning Fourier series, we take another approach. Shown in Figure 8.25 (solid line) is the aperture function $p_{r_c}(-\lambda d_2 u)$ and, as can be seen, it uniformly passes all frequencies up to its cutoff while completely removing all higher ones. This cutoff frequency is obviously given as

$$(8.59) \quad u_c = \frac{r_c}{\lambda d_2}.$$

Because the transform $\Psi_1(u, v)$ contains only discrete values that correspond to the Fourier series coefficients of the object $\psi_1(x_1, y_1)$, we see that the effect of the lens aperture function is to truncate the Fourier series. That is, the image $\psi_2(x_2, y_2)$ is described by the truncated Fourier series representation of the object. The number of terms in the series can be obtained by setting

$$N\Delta w = \frac{N}{4a} = \frac{r_c}{\lambda d_2}$$

which implies

$$(8.60) \quad N = \frac{4ar_c}{\lambda d_2}.$$

Since the Fourier series (rectangular form) of the object is

$$(8.61) \quad \psi_1(x_1, y_1) = \frac{1}{2} + \sum_{k=1}^{\infty} \text{sinc}\frac{\pi k}{2} \cos\frac{k\pi x}{2a},$$

the image must be described by the truncated Fourier series

$$(8.62) \quad \psi_2(x_2, y_2) = \frac{1}{2} + \sum_{k=1}^{N} \text{sinc}\frac{\pi k}{2} \cos\frac{k\pi x}{2a}.$$

From Chapter 2 we know that when we approximate a function with a truncated Fourier series we will always obtain ringing, or Gibbs' effect, at the discontinuities. Therefore, the image of our line pairs will exhibit "fringes" at their edges. A typical real world situation might be $a = 0.025$ cm, $d_2 = 100$ cm, $r_c = 3$ cm, $\lambda = 0.6238 \times 10^{-4}$ cm. Thus using Equation (8.60) we obtain

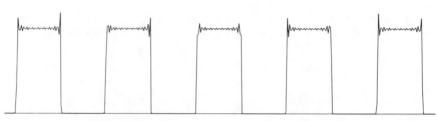

FIGURE 8.26.   Image of line pairs using 47 terms.

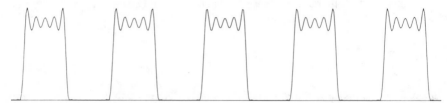

FIGURE 8.27.   Image of line pairs using nine terms.

$N = 47$ and Equation (8.62) will contain 47 terms of the Fourier series approximation. This is shown (modulus squared) in Figure 8.26. If we now change the spacing a from 0.025 cm to 0.005 cm, we find the Fourier series contains only nine terms. The resulting image is shown (modulus squared) in Figure 8.27.

COHERENT AND INCOHERENT OPTICAL SYSTEMS

In the previous section we presented a mathematical description of image formation using a single positive lens. Our analysis showed how we could consider a lens to be a system that converted an object to an image and we derived simple expressions for both the point spread and transfer function of this system. Although not specifically pointed out, this presentation assumed that the illumination was monochromatic. As a matter of fact, for almost all of our work in this chapter, we have made the assumption that the optical disturbance function was monochromatic [see Equation (8.8)]. In this section we illustrate how a general optical system (which produces a real image) can be modeled as a single positive lens. We also heuristically discuss imaging when the illumination (optical disturbance function) is either coherent or incoherent.

We assume that the purpose of an ideal optical system is to produce a real image of an object. Figure 8.28 shows a schematic diagram of such a system. In this figure we have denoted the distance between the object and image plane as d. We have also shown the input angle α_1 as well as the output angle α_2. It turns out that these three parameters are sufficient to describe the operation of

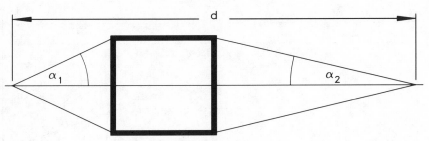

FIGURE 8.28.   Ideal imaging system.

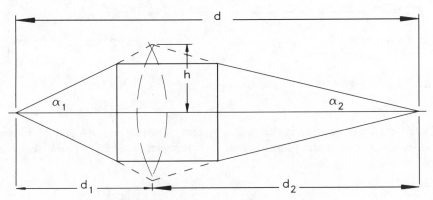

FIGURE 8.29.   Single lens model of imaging system.

this imaging system. Although they certainly depend upon the particular lens configuration within the system, they are measured externally and there are usually many different lens combinations that will yield the same parameters $\langle d, \alpha_1, \alpha_2 \rangle$, the simplest one obviously being a single positive lens such as illustrated in Figure 8.29.

Based on the previous comments, it should be obvious that we can model any imaging system in terms of a single positive lens.[†] This means that the analysis of the previous section can be used to describe any imaging system. That is, given any imaging system we can convert it to a single lens model and then use Equations (8.54) and (8.55) to determine its point spread and transfer functions. The geometrical reltaionships shown in Figure 8.29 are described mathematically as follows.

$$(8.63) \qquad\qquad d_1 + d_2 = d,$$

$$(8.64) \qquad\qquad h = d_1 \tan \alpha_1 = d_2 \tan \alpha_2.$$

[†]A single lens will always produce an inverted image whereas some optical systems produce an upright one. A simple sign change rectifies this problem.

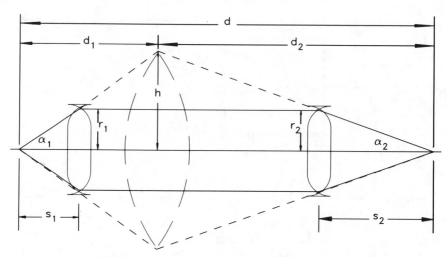

FIGURE 8.30. Example of imaging system single lens model.

As an example, we convert the optical system shown in Figure 8.30 (solid line) to its single lens model (dashed line). For this system let us suppose that we have the following parameters.

Lens 1	Lens 2
$r_1 = 5$ cm	$r_2 = 5$ cm
$s_1 = 30$ cm	$s_2 = 60$ cm
$f_1 = 30$ cm	$f_2 = 30$ cm

$$d = 300 \text{ cm.}$$

We first note

$$\tan \alpha_1 = \frac{r_1}{s_1} = \frac{5}{30},$$

$$\tan \alpha_2 = \frac{r_2}{s_2} = \frac{5}{60}.$$

Using Equation (8.64) we determine $d_2 = 2d_1$. The using Equation (8.63) we have $d_1 = 100$ cm and $d_2 = 200$ cm. Finally (using Equation 8.64 again), we find $h = 16.67$ cm. Therefore, our single lens model has a lens aperture function, or pupil function, that is described by

$$A_L(x, y) = \begin{cases} 1, & x^2 + y^2 \leqslant (16.67)^2 \\ 0, & \text{otherwise.} \end{cases}$$

The focal length of the single lens model is

$$f = \frac{d_1 d_2}{d_1 + d_2} = 66.67 \text{ cm}.$$

Now using Equation (8.54) we see that the point spread function is an Airy pattern whose central radius is

$$r = \frac{3.8 \lambda d_2}{2 \pi h} = 4.6 \times 10^{-4} \text{ cm} \quad \left(\text{assuming } \lambda = 0.6328 \times 10^{-4} \text{ cm}\right).$$

The transfer function of this system [Equation (8.55)] is a circ function with cutoff frequencies

$$u_{\max} = v_{\max} = \frac{h}{\lambda d_2} = 1317.2 \text{ cm}^{-1}.$$

Although we have not specifically mentioned it, up to this point in the chapter we have been dealing with coherent radiation. To appreciate this fact we must return to the scalar wave equation (8.6) and consider its general solution $V(\mathbf{x};t)$. We say that the radiation is (spatially) *coherent* if the time variation of $V(\mathbf{x};t)$—(for all space)—is uniform in time. Although no light is completely coherent, the output from a laser and light originating from a point source are examples of strongly coherent radiation. On the other hand we say the radiation is (spatially) *incoherent* if the time variation of $V(\mathbf{x};t)$ is completely random. Even though much of the optical radiation lies somewhere between these two extremes, it turns out that a study of coherent imaging systems and incoherent imaging systems is sufficient to handle most real world problems. We may recall that at the beginning of this chapter we assumed that $V(x;t)$ was a separable function in time and space [Equation (8.7)]. That is, we assumed that it could be described as the product of a pure time function and a pure space function. Based on out previous definitions this obviously implies coherent radiation.[†] As it turns out, we only require a slight modification in our previous results (for coherent systems) to yield similar results for the incoherent case. Before we begin our presentation, we emphasize the fact that the only difference between a coherent and incoherent optical system is the type of radiation (or illumination) used. In other words, the same physical lens configuration may be considered coherent or incoherent. It depends only on the object illumination.

We begin with an optical system that images a point source of light to the point spread function $h(x, y)$. Because radiation from a single point source is

---

[†]Actually, we even restricted ourselves further with Equation (8.8) when we assumed monochromatic radiation. However, the important thing to note here is that it was at Equation (8.7) that we made the assumption that the radiation was coherent.

coherent, we are able to use the results of the previous section (namely, Equation 8.54) to determine $h(x, y)$. Next we examine the imaging of two separate point sources. When the illumination is coherent, the point sources will vary in unison and, consequently, so too will the resulting impulse responses $h_1(x, y)$ and $h_2(x, y)$. Therefore, they must be added on a complex amplitude basis. That is, the image is given as

$$\text{Image}_c(x, y) = \|h_1(x, y) + h_2(x, y)\|^2.$$

When the illumination is incoherent, then the two point sources vary randomly with respect to each other and, consequently, their impulse responses also vary in some statistically independent fashion. Therefore, they must be added on a power or intensity basis. That is, the image is given as

$$\text{Image}_i(x, y) = \|h_1(x, y)\|^2 + \|h_2(x, y)\|^2.$$

Recall (Chapter 6) that when we convolve the input to a system with the system impulse response (to obtain the output) we are, in effect, considering the input (object) to be made up of impulses and the output to be the sum, or linear combination, of the resulting impulse responses. Thus based on our previous comments, we see for coherent illumination we convolve the object disturbance function with the point spread function on a complex amplitude basis:

$$(8.65) \qquad \psi_2(x_2, y_2) = h(x_2, y_2) * \psi_1(x_1, y_1).$$

However, when the illumination is incoherent we must convolve them on an intensity basis;

$$(8.66) \qquad \|\psi_2(x_2, y_2)\|^2 = \|h(x_2, y_2)\|^2 * \|\psi_1(x_1, y_1)\|^2.$$

Equations (8.65) and (8.66) express the differences (or similarities) between coherent and incoherent imaging systems. We note the relationship between the point spread functions. Namely, the incoherent point spread function is the modulus squared of the coherent spread function. Inasmuch as the transfer function of any system is simply the Fourier transform of its impulse response, it follows that the transfer function of an incoherent optical system is

$$H_I(u, v) = \mathcal{F}\left[\|h(x, y)\|^2\right] = \mathcal{F}\left[h(x, y)h^*(x, y)\right]$$

where $h^*(x, y)$ is the complex conjugate of $h(x, y)$. Using the product Theorem 3.10 and Theorem 3.13 for the transform of the complex conjugate we obtain

$$H_I(u, v) = \mathcal{F}\left[h(x, y)\right] * \mathcal{F}\left[h^*(x, y)\right] = H(u, v) * H^*(-u, -v).$$

However, $H(u, v) * H^*(-u, -v)$ is simply the autocorrelation of $H(u, v)$ with itself (see Chapter 3) and thus

(8.67) $$H_I(u, v) = H(u, v) \star H(u, v).$$

The preceding equation tells us that the transfer function of the incoherent optical system is equal to the transfer function of the coherent system correlated with itself. It is customary to normalize $H_I(u, v)$ by its DC value, that is, $H_I(0, 0)$. The result of this normalization is denoted as $\mathcal{K}(u, v)$ and called the *optical transfer function*;

(8.68) $$\mathcal{K}(u, v) = \frac{H(u, v) \star H(u, v)}{\|H(0, 0)\|^2}.$$

The modulus of this is known as the *modulation transfer function* or MTF.

We now present a graphical interpretation of Equation (8.68) for the case when the lens aperture, or pupil function, is real and symmetrical. Based on previous remarks, we know that the coherent transfer function $H(u, v)$ is just the pupil function scaled by $-\lambda d_2$. Let us denote the pupil function as $P(x, y)$. Thus using Equation (8.55) and taking advantage of our real and symmetrical assumptions we find

$$\mathcal{K}(u, v) = \frac{P(\lambda d_2 u, \lambda d_2 v) \star P(\lambda d_2 u, \lambda d_2 v)}{\|P(0, 0)\|^2}.$$

In integral form, the numerator of the preceding equation is

$$\int\int_{-\infty}^{\infty} P(\hat{\xi}, \hat{\eta}) P(\hat{\xi} + \lambda d_2 u, \hat{\eta} + \lambda d_2 v) \, d\hat{\xi} \, d\hat{\eta}.$$

Now making the change of (dummy) variables $\xi = \hat{\xi} + \lambda d_2 u/2$ and $\eta = \hat{\eta} + \lambda d_2 v/2$ we have

$$\int\int_{-\infty}^{\infty} P\left(\xi + \frac{\lambda d_2 u}{2}, \eta + \frac{\lambda d_2 v}{2}\right) P\left(\xi - \frac{\lambda d_2 u}{2}, \eta - \frac{\lambda d_2 v}{2}\right) d\xi \, d\eta.$$

We also point out that

$$\|P(0, 0)\|^2 = \int\int_{-\infty}^{\infty} P(\xi, \eta) P(\xi, \eta) \, d\xi \, d\eta = \int\int_{-\infty}^{\infty} P(\xi, \eta) \, d\xi \, d\eta.$$

Note that in the preceding equation, since the value of $P(\xi, \eta)$ is either 1 or 0 we have $P(\xi, \eta)P(\xi, \eta) = P(\xi, \eta)$. When all these moves are substituted into

Equation (8.68) we find

(8.69) $\mathcal{K}(u, v)$

$$= \frac{\iint P(\xi + \lambda d_2 u/2, \eta + \lambda d_2 v/2) P(\xi - \lambda d_2 u/2, \eta - \lambda d_2 v/2) \, d\xi \, d\eta}{\iint P(\xi, \eta) \, d\xi \, d\eta}$$

Examining this equation we see that, for any values of u and v, the numerator represents the area of overlap of two displaced pupil functions. One is centered at $(\lambda d_2 u/2, \lambda d_2 v/2)$ and the other is located at the diametrically opposite point $(-\lambda d_2 u/2, -d\lambda d_2 v/2)$. This is illustrated in Figure 8.31. The denominator is simply the total area of the pupil function. Thus we see the optical transfer function (for any frequencies u and v) is equal to the area of overlap of the displaced pupil functions normalized by the total area of the pupil function.

Our final topic in this section is a study of the incoherent imaging of the line pairs shown in Figure 8.24. Our optical system consists of a single lens with aperture function $A_L(x, y)$. Since we are considering the system to be incoherently illuminated, we must work with the intensity of the disturbance functions. We first note that the value of the object disturbance function $\psi_1(x_1, y_1)$ is either 1 or 0 and, therefore, we have

$$\|\psi_1(x_1, y_1)\|^2 = \psi_1(x_1, y_1) = p_a(x_1) * \mathrm{comb}_{4a}(x_1).$$

The Fourier transform of this expression [see Equation (8.57)] is

$$\mathcal{F}\left[\|\psi_1(x_1, y_1)\|^2\right] = 2a \, \mathrm{sinc}(2\pi u a) \mathrm{comb}_{1/4a}(u).$$

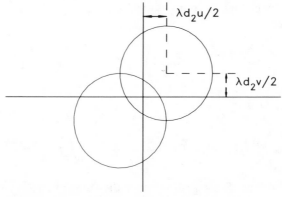

FIGURE 8.31.   Graphical interpretation of Equation (8.68).

As we saw in the previous section, this is simply the Fourier series representation of the object. When the object illumination was coherent, we learned that the transfer function of the lens had a sharp cutoff frequency and thus the image was a truncated Fourier series (see Figures 8.26 and 8.27). We now show that the transfer function of the incoherent optical system is a triangle function and thus gradually truncates the Fourier series. Just as we did in the coherent case, we assume that our lens is circular but because of the one-dimensional nature of our problem, we may consider the lens aperture function to be a pulse function:

$$A_L(x, y) = p_{r_c}(x).$$

Now using Equation (8.67) we find the incoherent transfer function to be

$$H_I(u, v) = p_{r_c}(-\lambda d_2 x) \star p_{r_c}(-\lambda d_2 x).$$

Because $p_{r_c}(x)$ is a real and even function, the preceding equation becomes a convolution product. In Chapter 3 (Figure 3.12) we determined that this particular convolution product was a triangle function with height $2r_c/\lambda d_2$ and base $4r_c/\lambda d_2$ as shown in Figure 8.32. From this figure we see that this incoherent system linearly reduces the Fourier series to zero at the cutoff frequency $2r_c/\lambda d_2$. We note that this cutoff frequency is twice that of the corresponding coherent system. Using analogous reasoning to that of the previous section [Equation (8.60)], we determine the number of terms in the truncated Fourier series to be

(8.70)
$$N_I = \frac{8ar_c}{\lambda d_2}.$$

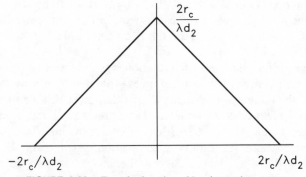

FIGURE 8.32. Transfer function of incoherent lens system.

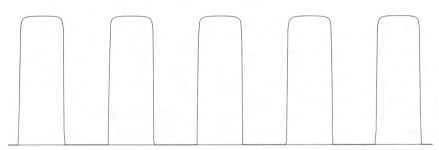

FIGURE 8.33. Incoherent image of line pairs using 94 terms.

The resulting image is given by the truncated Fourier series

$$(8.71) \quad \|\psi_2(x_2, y_2)\|^2 = \frac{1}{2} + \sum_{k=1}^{N} \left(a - \frac{N-k}{N} \right) \operatorname{sinc} \frac{\pi k}{2} \cos \frac{k\pi x}{2a}.$$

If, as in the previous section, we choose the situation $a = 0.025$ cm, $d_2 = 100$ cm, $r_c = 3$ cm, and $\lambda = 0.6328 \times 10^{-4}$ cm, we find $N_I = 94$ terms. The results of using Equation (8.71) with 94 terms is shown in Figure 8.33. This figure should be compared to Figure 8.26 which shows the same image for coherent illumination. As can be appreciated by comparing these two figures, the effect of linearly reducing, or truncating, the Fourier series (incoherent illumination) is to drastically reduce the "ringing" or fringes at the edges.

RESOLUTION OF OPTICAL SYSTEMS

In this section we examine the resolution of an optical system. Intuitively, we consider the resolution of an optical system to be a measure of how good an image it produces. Unfortunately, the quality of the image is strongly dependent upon the object and, therefore, it is very difficult to present a universal measure of the resolution of an optical system. In this section we consider the resolution of a system in terms of both its point spread function and its transfer function.

One of the earliest measures of resolution was how well an optical system could distinguish between two point sources separated by a distance $2a$. This was useful in the design of telescope systems that must be able to distinguish between two adjacent stars. As we learned in a previous section, a point source will be imaged to an Airy pattern (this, of course, assumes a circular pupil function). Furthermore, for a unity magnification system, two adjacent point sources separated by a distance $2a$ will be imaged to two adjacent Airy patterns also separated by a distance $2a$. This is illustrated graphically in one

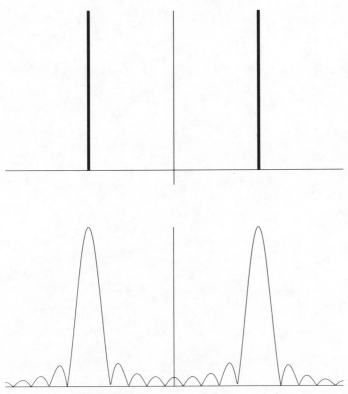

FIGURE 8.34. Two point resolution criterion.

dimension in Figure 8.34. Depending upon the width of the Airy patterns and their separation $2a$, these two images may overlap. Figure 8.35 shows various degrees of overlap between these images (in this figure we have assumed incoherent illumination). In Figure 8.35a we have no trouble in determining that this is the image of two distinct point source objects. However, in Figure 8.35c it is virtually impossible to determine if this is the image of one or two objects.

The task before us is to determine at what point the image overlap prohibits us from resolving the two distinct point objects. Although there are several such criteria available, we consider only the *Rayleigh criterion* for incoherent illumination here. This states that two points are resolved when the central maximum of one Airy pattern coincides with the first zero of the other. (this situation is illustrated in Figure 8.35b). We learned earlier in the chapter that the first zero of an Airy pattern was located at

$$\frac{3.8\lambda d_2}{2\pi b} = 1.22\frac{\lambda d_2}{2b}.$$

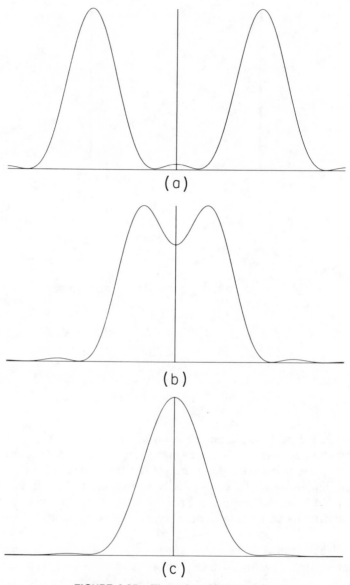

(a)

(b)

(c)

FIGURE 8.35.   Illustration of image overlap.

Recall that b is the radius of the lens and d_2 is the distance from the lens (or exit pupil) to the image plane. Thus we see that the separation $2a$ must be

$$(8.72) \qquad\qquad\qquad 2a = 1.22\frac{\lambda d_2}{2b}.$$

This equation gives us the "resolving power" of an incoherent optical system as per the Rayleigh criterion. As an example, let $\lambda = 0.6328 \times 10^{-4}$ cm, $d_2 = 30$ cm, $b = 5$ cm; then our system can resolve two point sources separated by a distance

$$2a = \frac{1.22(0.6328 \times 10^{-4})30}{10} = 2.3 \times 10^{-4} \text{ cm.}$$

We noted that the Rayleigh criterion is valid for incoherent illumination and does not apply when the illumination is coherent. We illustrate this as follows: let

$$A(r) = \frac{J_1(2\pi r/\lambda d_2)}{r/\lambda d_2}$$

be the Airy pattern (or image disturbance) of each point source. When the illumination is incoherent, the image is described as

$$\|\psi_2(r)\|^2 = \|A(r + a)\|^2 + \|A(r - a)\|^2.$$

However, when the illumination is coherent, the image is given as

$$\|\psi_2(r)\|^2 = \|A(r + a) + A(r - a)\|^2.$$

If we use the Rayleigh separation [Equation (8.72)] for this case, we find that there is no dip in the center of the image (see Figure 8.36). To give a satisfactory presentation of two-point resolution for the coherent case requires a knowledge of partial coherence and, therefore, we refer the interested reader to the text by Thompson and/or Goodman listed in the Bibliography at the end of this chapter. Heuristically, we note that even though the Rayleigh criterion is not valid for coherent illumination, it is often used anyway just to obtain a "ballpark" answer.

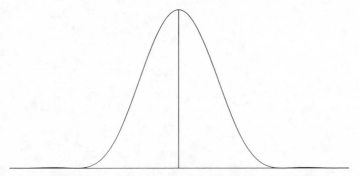

FIGURE 8.36. Rayleigh separation with coherent illumination.

Another common method of specifying the resolution of an optical system is based on how well it will image a bar chart, or set of equally spaced line pairs. Referring to Figure 8.24, we define the period of a bar chart as $4a$. However, it is more common to characterize a bar chart as the number of line pairs per mm. Thus we express a in mm and use $1/4a$ as the characteristic quantity. For example, if $4a = 0.01$ mm, then we would describe this bar chart as 100 lines per mm.

In previous sections we have discussed both the coherent and incoherent imaging of a set of line pairs. We showed that, for both types of illumination, the image was given as the truncated Fourier series representation of the object. The number of terms in the series (see Equations 8.60 and 8.70) is determined by the various physical dimensions of both the optical system (λ, d_2, r_c) and the bar chart ($4a$). Obviously, the more terms in the series, the "better" the image. We must now decide on how many terms are required before we can consider the object to be resolved. Clearly, if only the DC (zero frequency) coefficient is preserved, there is no image. However, if the first coefficient (above DC) is passed, we will have an image as illustrated in Figure 8.37. Clearly, this is a poor image of the original object. However, it is sufficient to determine the basic period of the bar chart. Therefore, let us agree that we can resolve a bar chart if the first nonzero frequency component is passed.

For a coherent system we have a definite cutoff frequency as given by Equation (8.59). This cutoff frequency can be used to determine the number of terms in the Fourer series as per Equation (8.60). Therefore, if in this equation we set $N = 1$, we find

$$(8.73) \qquad\qquad \frac{1}{4a} = \frac{r_c}{\lambda d_2},$$

which is the number of line pairs per mm that a coherent optical system will resolve. For example, if $r_c = 4$ cm, $\lambda = 0.6328 \times 10^{-4}$ cm, $d_2 = 100$ cm, then Equation (8.73) implies that we can resolve 632 line pairs per mm.

FIGURE 8.37. Minimum resolution of bar chart.

Specifying the resolution of an incoherent system (in terms of bar charts) is somewhat more nebulous than that of a coherent system. This fact can be appreciated by examining Figure 8.25 which shows the transfer function of a coherent system. As can be seen, this system uniformly passes all frequencies up to its cutoff. Thus the maximum resolution of a bar chart (with coherent illumination) is obtained when the first frequency coefficient (above DC) is located at (or just within) this cutoff value. An incoherent system, on the other hand, does not uniformly pass all frequencies within its pass-band as can be appreciated from Figure 8.32. The "amount" of the first frequency coefficient strongly depends upon its location in the frequency spectrum relative to the cutoff value. Thus when we specify the resolution of an incoherent system, we must give both the cutoff frequency and a contrast ratio, defined as

$$(8.74) \qquad C_r = \frac{I_{max} - I_{min}}{I_{max} + I_{min}}.$$

I_{max} and I_{min} are the maximum and minimum intensity values of the image. For our work here, we are only interested in using the first coefficient ($N_I = 1$) of Equation (8.71). Thus

$$I(x, y) = 0.5 + A \cos \frac{\pi x}{2a}$$

and $I_{max} = 0.5 + A$, $I_{min} = 0.5 - A$. Equation (8.74) then yields $C_r = 2A$ where $2A$ is the peak-to-peak amplitude of the first cosine term (above DC). We note [Equations (8.60) and (8.70)] that the cutoff frequency of the incoherent system is twice that of the coherent one and, at first glance, we may be tempted to conclude that it has twice the resolving power. However, at (or just within) the cutoff frequency the contrast ratio of the incoherent system is essentially zero, resulting in no image at all. On the other hand, if we locate the first coefficient midway within the interval ($0.2r_c/\lambda d_2$), then we see that we will indeed have a finite contrast ratio (Figure 8.32). In this case, both the coherent and incoherent systems may be considered to have the same resolution. The point to be made here is that the resolution of an incoherent system must be specified in terms of both the periodicity of the bar chart and a contrast ratio. For real world systems the proper (required) contrast ratio will depend upon the amount of noise or other degrading effects present.

In this section we have presented the concept of optical resolution in terms of both the point spread and transfer function of the system. For two point sources, resolution was defined in terms of the overlap of the point spread functions. For a bar chart we defined it in terms of the transfer function of the system. We did this by determining the minimum frequency content of the object that the system's transfer function must pass in order to resolve the object. This technique can often be generalized to an arbitrary object if we can somehow specify the minimum frequency content of the object required to resolve it. That is, the transfer function of the system must pass this minimum frequency.

SPATIAL FILTERING

In this section we discuss the special two-lens imaging system shown in Figure 8.38. For reasons that will soon become apparent, this system is known as a spatial filtering optical system. In this system we have two identical lenses (of focal length f) located as shown in the figure. As we show, this system produces an (inverted) image at plane 3 of an object located at plane 1. The unique feature of this system is that at plane 2 (midway between the lenses), the optical disturbance is equal to the Fourier transform of the object's optical disturbance function. In other words, as the radiation from the object propagates through the system it experiences diffraction effects. At plane 2 the resulting diffraction pattern will be the Fourier transform of the object. As the radiation propagates past this plane the second lens modifies the diffraction in such a way that an image results at plane 3.

The frequency content of the object is spatially displayed in plane 2 and, consequently, we have the opportunity to remove or filter various portions of it. This obviously results in a modified image at plane 3. At the end of this section we present several examples illustrating the results of spatial filtering.

We now show that the optical disturbance at plane 2 is the Fourier transform of the optical disturbance at plane 1. Let us denote the disturbance just before lens 1 as $\psi_-(x, y)$ and that just after the lens as $\psi(x, y)$. We can use Equation (8.21) to propagate $\psi(x, y)$ the distance f to plane 2;

$$\psi_2(x_2, y_2) = \frac{K}{f} e^{ikf} e^{(ik/2f)(x_2^2 + y_2^2)} \mathcal{F}_s \left[\psi(x, y) e^{(ik/2f)(x^2 + y^2)} \right].$$

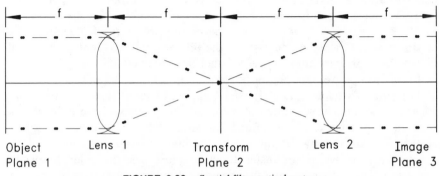

FIGURE 8.38. Spatial filter optical system.

From the lens Equation (8.36) we obtain

$$\psi(x, y) = L\psi_-(x, y)e^{-(ik/2f)(x^2+y^2)}.$$

When this is substituted into the previous equation, the quadratic phase terms cancel and we obtain

(8.75)
$$\psi_2(x_2, y_2) = \frac{KL}{f}e^{ikf}e^{(ik/2f)(x_2^2+y_2^2)}\mathscr{F}_s[\psi_-(x, y)].$$

Now using Equation (8.29) we can determine the Fourier transform of $\psi_-(x, y)$ in terms of the Fourier transform of the disturbance at plane 1 $\psi_1(x_1, y_1)$:

$$\mathscr{F}[\psi_-(x, y)] = e^{ikf}\mathscr{F}[\psi_1(x_1, y_1)]e^{-i\pi\lambda f(u^2+v^2)}.$$

Taking into account the scale factors (i.e., $u = x_2/\lambda f$ and $v = y_2/\lambda f$) in this equation we find

$$\mathscr{F}_s[\psi_-(x, y)] = e^{ikf}\mathscr{F}_s[\psi_1(x_1, y_1)]e^{-(ik/2f)(x_2^2+y_2^2)}.$$

Finally, substituting this expression back into Equation (8.75) and cancelling the quadratic phase terms we arrive at

(8.76)
$$\psi_2(x_2, y_2) = \frac{KL}{f}e^{i2kf}\mathscr{F}_s[\psi_1(x_1, y_1)].$$

which is our desired result. Using an identical approach we can also show that

$$\psi_3(x_3, y_3) = \frac{KL}{f}e^{i2kf}\mathscr{F}_s[\psi_2(x_2, y_2)].$$

Now, using Equation (8.76) and the transform of a transform theorem 3.20 we obtain

(8.77)
$$\psi_3(x_3, y_3) = \frac{K^2L^2}{f^2}e^{i4kf}\psi_1(-x_1, -y_1)$$

or, in other words, $\psi_3(x_3, y_3)$ is an inverted image of the object $\psi_1(x_1, y_1)$.

As an illustration of spatial filtering, we offer the following examples. Shown in Figure 8.39a is a bar chart. Its optically obtained Fourier transform is shown in Figure 8.39b (recall that this situation was analyzed mathematically in a previous section and these figures should be compared to those of Figures 8.24 and 8.25). In Figure 8.40a we show the transform plane in which all but

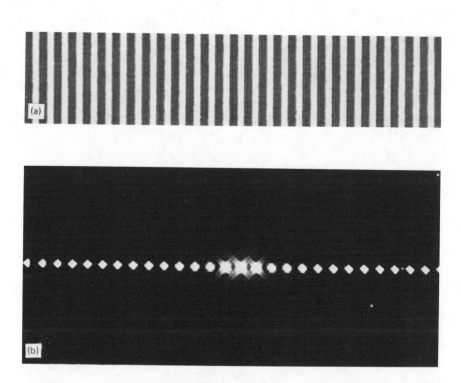

FIGURE 8.39. (*a*) Bar chart. (*b*) Optically obtained Fourier transform.

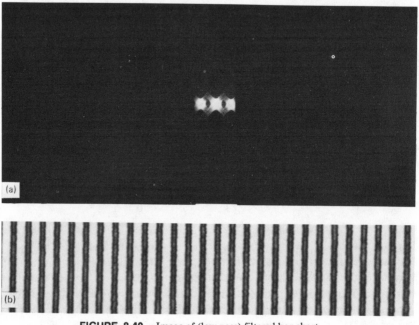

FIGURE 8.40. Image of (low pass) filtered bar chart.

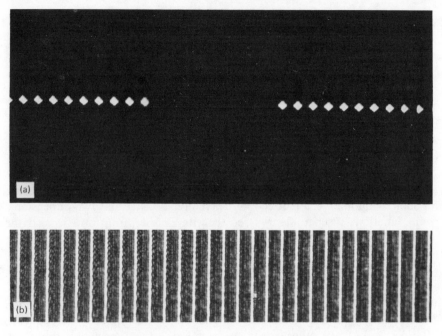

FIGURE 8.41. Image of (high pass), filtered bar chart.

the first coefficient (or frequency above DC) have been removed. The resulting image (plane 3) is shown in Figure 8.40b. The next filter situation is shown in Figure 8.41a in which the first three coefficients (above DC) are removed and only the higher frequencies passed. The resulting image is shown in Figure 8.41b.

The purpose of these previous examples has been to illustrate the principles of spatial filtering and, consequently, they were quite simple. However, using holographic techniques, complex filters can be constructed that selectively modify both the amplitude and phase of the transform at plane 2. For more information concerning these filters, the reader is referred to the text by Goodman listed in the Bibliography at the end of this chapter. In the next section we combined spatial filtering and sampling theory to obtain some rather impressive results.

SAMPLING THEORY WITH OPTICS APPLICATIONS

In this section we present an optical illustration of sampling theory. We begin by generalizing our comb function approach to sampling to two dimensions.

We define the two-dimensional comb function as

$$(8.78) \quad \text{comb}_{\Delta x, \Delta y}(x, y) = \sum_{k_1=-\infty}^{\infty} \sum_{k_2=-\infty}^{\infty} \delta(x - k_1\Delta x)\delta(y - k_2\Delta y).$$

This function is obviously separable and can be written as the product of two one-dimensional comb functions:

$$(8.79) \qquad \text{comb}_{\Delta x, \Delta y}(x, y) = \text{comb}_{\Delta x}(x)\text{comb}_{\Delta y}(y).$$

As we learned in Chapter 5, the Fourier transform of a one-dimensional comb function is another comb function. Therefore, because the two-dimensional comb function is separable, we see that its Fourier transform is also the product of two comb functions:

$$(8.80) \qquad \mathcal{F}\left[\text{comb}_{\Delta x, \Delta y}(x, y)\right] = \text{comb}_{1/\Delta x}(u)\text{comb}_{1/\Delta y}(v).$$

When we multiply a two-dimensional function $f(x, y)$ by the comb function $\text{comb}_{\Delta x, \Delta y}(x, y)$, we convert that function to a two-dimensional sequence $\{f(k_1\Delta x, k_2\Delta y)\}$. In other words, we sample the function with a uniformly spaced interval $(\Delta x, \Delta y)$. Let us assume that $f(x, y)$ is bandlimited with bandwidth $(2\Omega_1, 2\Omega_2)$. Furthermore, let us denote the sampled sequence as

$$h(x, y) = \{f(k_1\Delta x, k_2\Delta y)\} = f(x, y)\text{comb}_{\Delta x}(x)\text{comb}_{\Delta y}(y).$$

Taking the Fourier transform of the preceding equation and using the product Theorem 3.22 we obtain

$$H(u, v) = F(u, v) * \text{comb}_{1/\Delta x}(u)\text{comb}_{1/\Delta y}(v).$$

However, since the delta function is the unit element under convolution, this equation becomes

$$(8.81) \qquad H(u, v) = \sum_{j_1=-\infty}^{\infty} \sum_{j_2=-\infty}^{\infty} F\left(u - \frac{j_1}{\Delta x}, v - \frac{j_2}{\Delta y}\right).$$

Equation (8.81) tells us that the Fourier transform of the sampled sequence is simply that of the original, or generating, function periodically displaced by the amounts $(1/\Delta x, 1/\Delta y)$. Therefore, if the sampling rate is equal to, or less than, the Nyquist rate (i.e., $1/\Delta x \leqslant 2\Omega_1$ and $1/\Delta y \leqslant 2\Omega_2$), then these transforms will not overlap and the original function $f(x, y)$ can be recoved by inverse Fourier transforming over the two-dimensional domain $[(-\Omega_1, \Omega_1) \times (-\Omega_2, \Omega_2)]$.

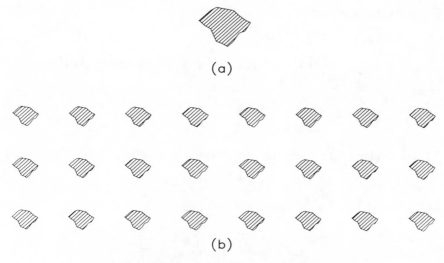

(a)

(b)

FIGURE 8.42. (*a*) Basic function. (*b*) Sampling array.

When we optically sample an object or function, we effectively do so by placing a mesh or periodic hole pattern over the object. When the individual holes are small, they may be considered points or delta distributions and thus our previous analysis is valid and can be used to describe optical sampling. However, we now generalize our previous analysis to the situation where the sampling is performed with a periodic array, or repetition, of a basic function. This situation is illustrated in Figure 8.42. Shown in Figure 8.42*a* is the basic function $a(x, y)$ and in Figure 8.42*b*, the periodic array generated by the function. The sampling array, denoted as $g(x, y)$ is described mathematically as

$$(8.82) \qquad g(x, y) = a(x, y) * \text{comb}_{\Delta x, \Delta y}(x, y).$$

Denoting the Fourier transform of $a(x, y)$ as $A(u, v)$ and using the convolution Theorem 3.21, we determine the Fourier transform of the sampling array to be given as

$$(8.83) \qquad G(u, v) = A(u, v)\text{comb}_{1/\Delta x}(u)\text{comb}_{1/\Delta y}(v).$$

When we use $g(x, y)$ to sample a function $f(x, y)$ we write

$$(8.84) \quad h(x, y) = f(x, y)g(x, y) = f(x, y)\big[a(x, y) * \text{comb}_{\Delta x, \Delta y}(x, y)\big].$$

Transforming Equation (8.84) and using the product Theorem 3.22, as well as

Equation (8.83), we see

$$H(u, v) = F(u, v) * G(u, v)$$

$$= \left\{ F(u, v) * \left[A(u, v) \, \text{comb}_{1/\Delta x}(u) \, \text{comb}_{1/\Delta y}(v) \right] \right\},$$

or

$$(8.85) \quad H(u, v) = \sum_{j_1 = -\infty}^{\infty} \sum_{j_2 = -\infty}^{\infty} F\left(u - \frac{j_1}{\Delta x}, v - \frac{j_2}{\Delta y} \right) A\left(\frac{j_1}{\Delta x}, \frac{j_2}{\Delta y} \right).$$

This is similar to the result that we obtained in Equation (8.81). That is, the

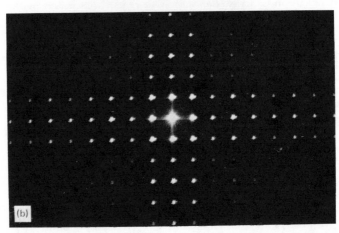

FIGURE 8.43. Optically recovered object from sampled distribution.

FIGURE 8.43. (*Continued*).

Fourier transform of the sampled distribution is that of the original function periodically displaced by the amounts $(1/\Delta x, 1/\Delta y)$. However, in this case each pattern is scaled by the *value* of the transform of the basic function at that location. Again we note that if the sampling rate is equal to or less than the Nyquist rate, then the transform patterns will not overlap and we are able to recover the original function $f(x, y)$.

We now illustrate this situation optically. Shown in Figure 8.43a is an object that has been sampled with an array of periodically spaced rectangular holes. This object is then placed in the spatial filter system of Figure 8.38. Shown in Figure 8.43b is the resulting optical disturbance at plane 2 or, in other words, the Fourier transform of the sampled object. As can be seen in this figure, and appreciated by our previous analysis, we have the Fourier transform of the original object periodically located throughout the plane. As shown in Figure 8.43c, we use a simple rectangular aperture to remove, or filter out, all but one such pattern. In plane 3 we then have the resulting image which is simply the (inverted) inverse Fourier transform of the filtered optical disturbance in plane 2 or, in this case, the original object. This is shown (inverted) in Figure 8.43d. Thus we have optically recovered the original function from its sampled distribution. The important thing to note in this demonstration is the removal of the sampling array. The (dust spot) diffraction rings that appear in Figure 8.43d are also in Figure 8.43a although they are somewhat obscured by the sampling pattern.

SUMMARY

In this chapter we first presented a physical derivation of the Fresnel diffraction equation from Maxwell's equations. This derivation was not intended to be rigorous but, instead, was presented so that the various limitations of the Fresnel equation could be appreciated. We then showed how the Fresnel equations could be written in terms of Fourier transforms. These results were used to study such topics as diffraction, imaging, resolution, and spatial filtering. These topics were presented using several examples to illustrate both the use and power of a Fourier analysis approach to optics. Entire books have been written that discuss the topics presented in this chapter. The interested reader is encouraged to examine the Bibliography list at the end of this chapter.

BIBLIOGRAPHY

Thompson, B. J., and G. B. Parrent, *Physical Optics Notebook*. S.P.I.E., Redondo Beach, Calif., 1969.

Goodman, J. W., *Introduction to Fourier Optics*. McGraw-Hill, New York, 1968.

Papoulis, A., *Systems and Transforms with Applications in Optics*. McGraw-Hill, New York, 1968.

Garbuny, M., *Optical Physics*. Academic, New York, 1965.

Sears, F. W., *Optics*. Addison-Wesley, Cambridge, Mass., 1949.

Born, M. and E. Wolf, *Principles of Optics*. 2nd rev. ed. Pergamon, New York, 1964.

Raven, F. H., *Mathematics of Engineering Systems*. McGraw-Hill, New York, 1966.

9

NUMERICAL ANALYSIS

In this chapter we take a look at the field of numerical analysis from a Fourier analysis point of view. We demonstrate that when numerical analysis algorithms are viewed in the frequency domain and considered as transfer functions, much insight into their behavior can be obtained. The approach that we take is to first present the general theory of considering an algorithm as a transfer function and then actually calculate the transfer functions for several of the more common and useful ones.

NONRECURSIVE ALGORITHMS

An *algorithm* is a rule by which we numerically operate on a function, or a set of data, to obtain some specified result. For example, to calculate the average of a set of scores, the algorithm would tell us to sum the scores and then divide by the total number of scores. Mathematically, a *nonrecursive algorithm* is described as

$$(9.1) \qquad y(k) = \sum_{l=-L}^{L} a_l f(k+l) \qquad k \in [0, N-1],$$

where $\{f(k)\}$ represents the original data or sampled function, and $\{y(k)\}$

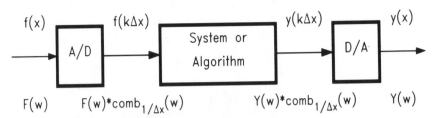

FIGURE 9.1. Numerical analysis problem.

represents the manipulated data. Algorithms are also known as digital filters or discrete systems.

Shown in Figure 9.1 is a schematic diagram of a numerical analysis problem in which a function $f(x)$ is to be numerically manipulated to yield a new function $y(x)$. The function $f(x)$ is first sampled and converted to a sequence. It is then operated on by the system as per Equation (9.1). Finally, it is converted back to a function with a D/A converter. Of concern to us here is the effect of the overall process of the frequency content of the original function $f(x)$. In other words, we want to determine the transfer function $H(w)$ of the system.

We demonstrate that the A/D and D/A converters (if chosen properly) have no effect on the frequency content of the function $f(x)$ and that the only modification comes from the transfer function of the algorithm. We begin by considering the transfer function of A/D or sampling operation. The input is a function $f(x)$ with frequency content or Fourier transform $F(w)$ and the output is the sequence $\{f(k)\}$. We learned in Chapter 5 that we could consider this output to be a function described by

$$f(x)\,\text{comb}_{\Delta x}(x).$$

The Fourier transform of this function is given as the following convolution product

$$F(w)*\text{comb}_{1/\Delta x}(w).$$

Note that using the comb function approach to sampling, $\{f(k\Delta x)\}$ may be considered either a sequence or a function (more properly a distribution). We pointed out in Chapter 5 that if the sampling rate Δx was chosen such that $\Delta x \leqslant 1/2\Omega$ [where 2Ω is the bandwidth of $f(x)$], then over the frequency interval $[-\Omega, \Omega]$,

$$F(w)*\text{comb}_{1/\Delta x}(w) = F(w) \qquad w \in [-\Omega, \Omega].$$

In this chapter we always implicitly assume that $\Delta x \leqslant 1/2\Omega$ and, therefore, the sampler can be considered to pass all the frequencies unchanged; that is, the transfer function is unity over the domain $[-\Omega, \Omega]$.

We now demonstrate that if the D/A converter uses the sampling theorem 5.1 interpolation rule when recovering $y(x)$ from $\{y(k)\}$, then all the frequency content of $y(x)$ will be preserved. We use a "back door" approach to show this result. That is, we first assume that all frequencies are uniformly passed and show that this results in the sampling theorem interpolation rule. We begin by noting that the frequency content of $\{y(k)\} = y(x)\,\mathrm{comb}_{\Delta x}(x)$ is given as

$$Y(w) * \mathrm{comb}_{1/\Delta x}(w).$$

Thus to pass all frequency content of $Y(w)$ (i.e., $-\Omega$ to Ω) we simply multiply this expression by a pulse function of half-width Ω. That is,

$$Y(w) = \left[Y(w) * \mathrm{comb}_{1/\Delta x}(w) \right] P_\Omega(w).$$

Now taking the inverse Fourier transform of both sides of the preceding equation and making use of the convolution and product theorems, we obtain

(9.2) $\qquad y(x) = \mathscr{F}^{-1}\left[Y(w) * \mathrm{comb}_{1/\Delta x}(w) \right] * \mathscr{F}^{-1}\left[P_\Omega(w) \right]$

However,

$$\mathscr{F}^{-1}\left[P_\Omega(w) \right] = \mathrm{sinc}\, 2\pi\Omega x,$$

$$\mathscr{F}^{-1}\left[Y(w) * \mathrm{comb}_{1/\Delta x}(w) \right] = y(x)\,\mathrm{comb}_{\Delta x}(x) = \sum_{k=-\infty}^{\infty} y(x)\delta(x - k\Delta x).$$

Thus Equation (9.2) becomes

$$y(x) = \left[\sum_{k=-\infty}^{\infty} y(x)\delta(x - k\Delta x) \right] * \mathrm{sinc}\, 2\pi\Omega x$$

$$= \sum_{k=-\infty}^{\infty} y(k\Delta x)\mathrm{sinc}(2\pi\Omega[x - k\Delta x]),$$

which is indeed the sampling theorem 5.1 interpolation rule.

Based on these discussions we can see that the only portion of the overall system with which we need concern ourselves is the transfer function of the algorithm.

Let us begin by rewriting Equation (9.1) in terms of the E operator defined as

$$E^l y(k) = y(k + l).$$

This results in

(9.3)
$$y(k) = \sum_{l=-L}^{L} a_l E^l f(k).$$

Next we consider the way in which the E operator works on the function

$$e(k) = e^{2\pi iwk\Delta x}.$$

That is,

$$E^l e(k) = e^{2\pi iw(k+l)\Delta x} = e^{2\pi iwl\Delta x} e^{2\pi iwk\Delta x} = e^{2\pi iwl\Delta x} e(k).$$

Therefore, if we run this function $e(t)$ through the system of Equation (9.3) we find

$$y(k) = \left(\sum_{l=-L}^{L} a_l e^{2\pi iwl\Delta x} \right) e^{2\pi iwk\Delta x}.$$

From this equation and the results of Chapter 6, we conclude that $\exp(2\pi iwk\Delta x)$ is an eigenfunction of this system and that

(9.4)
$$H(w) = \sum_{l=-L}^{L} a_l e^{2\pi iwl\Delta x},$$

is the system's transfer function. Thus our overall system can be described in the frequency domain as

(9.5)
$$Y(w) = H(w)F(w),$$

where $Y(w)$ is the Fourier transform of $y(x)$, $F(w)$ is the Fourier transform of $f(x)$, and $H(w)$ is the transfer function of the algorithm.

Before we leave this section, let us briefly consider the properties of linearity and invariance from a frequency domain point of view. We assume that $y_1(t)$ is the output of a system resulting from input $f_1(t)$ and, similarly, $y_2(t)$ is the output caused by $f_2(t)$. Also let us assume that their Fourier transforms are given by $Y_1(w)$, $F_1(w)$, $Y_2(w)$, and $F_2(w)$, respectively. Now as input to the system let us use a linear combination of both of the previous inputs; that is, $f(t) = af_1(t) + bf_2(t)$. In the frequency domain this becomes (Theorem 3.1)

$$F(w) = aF_1(w) + bF_2(w),$$

and using Equation (9.5), the resulting output is

$$Y(w) = H(w)F(w) = H(w)[aF_1(w) + bF_2(w)]$$

$$= aH(w)F_1(w) + bH(w)F_2(w)$$

$$= aY(w) + bY_2(w),$$

or back in the time domain $y(t) = ay_1(t) + by_2(t)$. This establishes linearity.

Now to look at invariance, let us use the function $f(t - \tau)$ which, in the frequency domain, is $F(w)\exp(-2\pi iw\tau)$. The resulting output (Equation 9.5) is

$$H(w)F(w)e^{-2\pi iw\tau} = Y(w)e^{-2\pi iw\tau},$$

or in the time domain, $y(t - \tau)$. This establishes invariance.

NONRECURSIVE SMOOTHING ALGORITHMS

In this section we look at the transfer function of several nonrecursive smoothing algorithms.

Smoothing by 3's

If in Equation (9.1) we let $L = 1$ and $a_{-1} = a_0 = a_1 = \frac{1}{3}$, then we obtain

(9.6) $$y(k) = \frac{1}{3}[f(k - 1) + f(k) + f(k + 1)].$$

Equation (9.6) is the classical "smoothing by 3's" nonrecursive digital filter. Using Equation (9.4) we can immediately write the transfer function as

$$H(w) = \frac{1}{3}(e^{-2\pi iw\Delta x} + 1 + e^{2\pi iw\Delta x}),$$

or

(9.7) $$H(w) = \frac{1 + 2\cos 2\pi w\Delta x}{3}.$$

This transfer function is plotted in Figure 9.2 from $w = 0$ to $w = w_{max} = 1/2\Delta x$.

At this time we point out certain important facts. First, because the sampling rate is Δx, we know (Chapter 5) that the highest possible frequency is

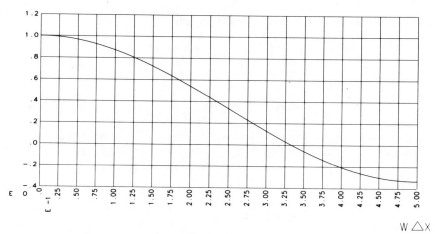

FIGURE 9.2. "Smoothing by 3's" transfer function.

$w_{max} = \Omega = 1/2\Delta x$. Second, for convenience the frequency scale of all our graphs is normalized by twice this maximum frequency. That is, we plot $w\Delta x$ (rather than w) from 0 to $\frac{1}{2}$.

Smoothing by 5's

The "smoothing by 5's" algorithm is obtained by letting $L = 2$ and $a_{-2} = a_{-1} = a_0 = a_1 = a_2 = \frac{1}{5}$ in Equation (9.1):

$$(9.8) \quad y(k) = \tfrac{1}{5}\left[f(k-2) + f(k-1) + f(k) + f(k+1) + f(k+2) \right].$$

Direct substitution into Equation (9.4) yields the transfer function

$$H(w) = \tfrac{1}{5}\left(e^{-4\pi i w\Delta x} + e^{-2\pi i w\Delta x} + 1 + e^{2\pi i w\Delta x} + e^{4\pi i w\Delta x} \right),$$

or

$$(9.9) \quad H(w) = \frac{1 + 2\cos 2\pi w\Delta x + 2\cos 4\pi w\Delta x}{5}.$$

This transfer function is shown in Figure 9.3.

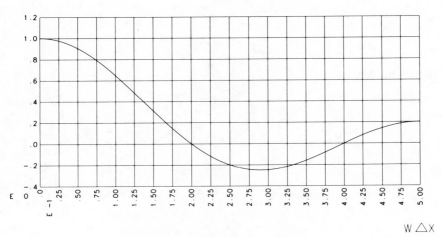

FIGURE 9.3. "Smoothing by 5's" transfer function.

Smoothing by (2L + 1)'s

In the general case, when we smooth by $(2L + 1)$'s, we set the coefficients all equal to $1/(2L + 1)$ and thus Equation (9.1) becomes

$$(9.10) \quad y(k) = \frac{1}{2L + 1} [f(k - L) + \cdots + f(k) + \cdots + f(k + L)].$$

Following the same procedure as before, we determine the transfer function to be

$$(9.11) \qquad H(w) = \frac{1 + 2\cos 2\pi w\Delta x + \cdots + 2\cos 2L\pi w\Delta x}{2L + 1}$$

or, more compactly,

$$(9.12) \qquad H(w) = \left(\frac{1}{2L + 1}\right) \frac{\cos 2\pi wL\Delta x + \cos 2\pi w(L + 1)\Delta x}{1 - \cos 2\pi w\Delta x}$$

This formula is plotted in Figure 9.4 for several values of $(2L + 1)$. As can be seen from this figure, the greater the value of L, the more rapidly the transfer function rolls off and the more severe the oscillations become.

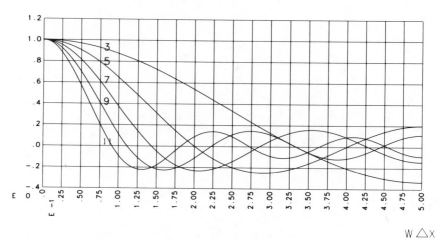

FIGURE 9.4. "Smoothing by $(2L + 1)$'s" transfer functions.

DERIVATIVE ALGORITHMS

In this section we consider the transfer function of algorithms that are used to estimate the derivative of a function. We begin our discussion by introducing the *first forward difference* of a function $\Delta f(x)$, defined as

$$\Delta f(x) = f(x + \Delta x) - f(x).$$

The first forward difference represents an incremental change in the function $f(x)$ corresponding to an incremental change in the variable x. This concept is quite useful to us in the remainder of this section. As an illustration of the first forward difference, let us now consider the following examples.

If $f(x) = x^2$, then

$$\Delta f(x) = (x + \Delta x)^2 - x^2 = 2x\Delta x + \Delta x^2.$$

$$\Delta \ln(x) = \ln(x + \Delta x) - \ln(x) = \ln(1 + \Delta x/x).$$

$$\Delta 3^x = 3^{x+\Delta x} - 3^x = 3^x(3^{\Delta x} - 1).$$

If c is constant, then

$$\Delta c = c - c = 0.$$

This operation can be repeated. That is, the *second forward difference* of a function is defined as

$$\Delta^2 f(x) = \Delta[\Delta f(x)] = \Delta[f(x + \Delta x) - f(x)]$$

$$= f(x + 2\Delta x) - 2f(x + \Delta x) + f(x).$$

In general, the *nth forward difference* is given as

$$\Delta^n f(x) = \Delta[\Delta^{n-1} f(x)] = \sum_{i=0}^{n} (-1)^i \binom{n}{i} f(x + n - i),$$

where

$$\binom{n}{i} = \frac{n!}{(n-i)! \, i!}.$$

The derivative of a function is defined as the limit

$$f'(x) = \lim_{\Delta x \to 0} \frac{f(x + \Delta x) - f(x)}{\Delta x} = \lim_{\Delta x \to 0} \frac{\Delta f(x)}{\Delta x}.$$

In numerical analysis we often approximate the derivative by the quotient $\Delta f(x)/\Delta x$. In terms of the sampled sequence we have

(9.13) $$y(k) = \frac{f(k+1) - f(k)}{\Delta x},$$

where it is understood that $f(k)$ means the function f evaluated at $x = k\Delta x$. This expression can be placed in the form of Equation (9.1) if we let $a_0 = -1/\Delta x$ and $a_1 = 1/\Delta x$. Thus we can use Equation (9.4) to obtain the transfer function

$$\frac{1}{\Delta x} (e^{2\pi i w \Delta x} - 1).$$

Now if we use the fact that

$$(e^{2\pi i w \Delta x} - 1) = e^{i \pi w \Delta x} (e^{i \pi w \Delta x} - e^{-i \pi w \Delta x}),$$

then we can rewrite this equation as

(9.14) $$2i e^{i \pi w \Delta x} \frac{\sin \pi w \Delta x}{\Delta x} = 2\pi i w e^{i \pi w \Delta x} \operatorname{sinc} \pi w \Delta x.$$

It is instructive to compare this transfer function to that of the actual derivative itself. Using theorem 3.8, we know that the Fourier transform of the derivative function is given as

$$Y(w) = \mathcal{F}[f'(x)] = 2\pi i w F(w),$$

and, therefore, its transfer function is

(9.15) $$H(w) = 2\pi i w.$$

Note that these transfer functions are both complex. That is, they consist of an amplitude and phase term. Shown in Figure 9.5 is an amplitude plot of both the algorithm transfer function (solid line) and the derivative transfer function (dashed line). (NOTE: The amplitude of each plot is normalized by Δx.)

Let us now normalize the algorithm transfer function by the derivative transfer function; that is, divide Equation (9.14) by Equation (9.15). This results in

(9.16) $$e^{i\pi w \Delta x} \operatorname{sinc} \pi w \Delta x.$$

This equation is a measure of the accuracy with which the algorithm approximates the derivative. Its amplitude is plotted in Figure 9.6. As can be appreciated from Equation (9.15), the effect of the derivative is to enhance the higher frequency components. This is also true for the derivative algorithm of Equation (9.13). However, it does not perform as well for the higher frequencies as can be seen by an examination of Figure 9.6.

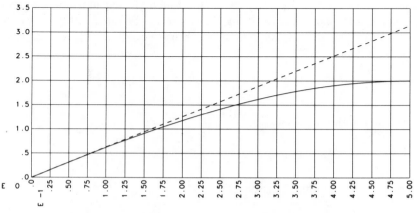

$$W \triangle X$$

FIGURE 9.5. First forward difference transfer function.

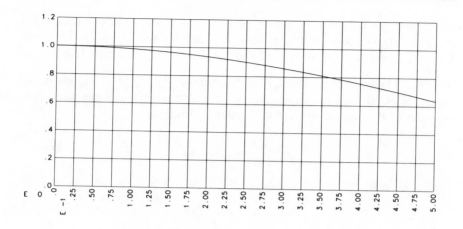

FIGURE 9.6. Accuracy of the first difference algorithm.

The second derivative can also be approximated using this type of approach:

$$(9.17) \qquad f''(x) \approx \frac{\Delta^2 f(x)}{\Delta x^2} = \frac{f(k + 2) - 2f(k + 1) + f(k)}{\Delta x^2}.$$

Again making use of Equations (9.1) and (9.4) (with $a_0 = 1/\Delta x^2$, $a_1 = -2/\Delta x^2$, and $a_2 = 1/\Delta x^2$), we obtain

$$(9.18) \qquad H(w) = \frac{1}{\Delta x^2}[e^{4\pi i w \Delta x} - 2e^{2\pi i w \Delta x} + 1].$$

This equation can be simplified by noting that the term in the brackets is simply

$$(e^{2\pi i w \Delta x} - 1)^2.$$

Using the same type of algebraic manipulation as before, the preceding equation becomes

$$(9.19) \qquad (2i)^2 e^{2\pi i w \Delta x}\left(\frac{\sin \pi w \Delta x}{\Delta x}\right)^2 = (2\pi i w)^2 e^{2\pi i w \Delta x}\mathrm{sinc}^2 \pi w \Delta x.$$

If we again use Theorem 3.8, we can show that the transfer function of the second derivative operation is

$$(9.20) \qquad H(w) = (2\pi i w)^2.$$

We have plotted the amplitude of both of these transfer functions [Equation (9.19) (solid), Equation (9.20) (dashed)] in Figure 9.7 and, as can be seen, both have the property of eliminating the low and enhancing the high frequencies.

NOTE: The amplitude of each plot is normalized by $(\Delta x)^2$. If we now divide Equation (9.19) by Equation (9.20), then we obtain a relative measure of the accuracy of this second forward difference algorithm:

$$(9.21) \qquad\qquad e^{2\pi i w\Delta x}\left(\text{sinc } \pi w\Delta x\right)^2$$

Figure 9.8 gives a plot of Equation (9.21) from which we can see that the

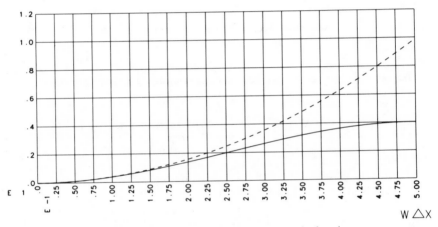

FIGURE 9.7. Second forward difference transfer function.

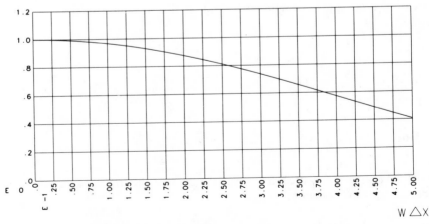

FIGURE 9.8. Accuracy of the second difference algorithm.

algorithm works well for low frequencies but deviates rather seriously for high frequencies.

This type of process can be continued to approximate the nth derivative of a function by using the nth forward difference of the function:

$$(9.22) \qquad f^{[n]}(x) \approx \frac{\Delta^n f(x)}{\Delta x^n} = \frac{f(k+n) + \cdots +/- f(k)}{\Delta x^n}.$$

Following the same procedure as before, it can be shown that the transfer function of this algorithm is

$$(9.23) \qquad H(w) = (2\pi i w)^n e^{in\pi w \Delta x} (\operatorname{sinc} \pi w \Delta x)^n$$

It can also be shown that the transfer function of the nth derivative operation is $(2\pi i w)^n$ and when this is used to normalize Equation (9.23), we obtain the following measure of the accuracy of the nth forward difference algorithm:

$$(9.24) \qquad e^{in\pi w \Delta x} (\operatorname{sinc} \pi w \Delta x)^n.$$

Another common method of numerically approximating the derivative of a function is by the *first central difference* operation, defined as

$$\Delta_c f(x) = f(x + \Delta x) - f(x - \Delta x).$$

Comparing this central difference formula to that of the first forward difference, we see that although the forward difference uses only values of the function forward of x, the central difference formula uses values of the function on both sides of x (but not at the value of x itself). We approximate the first derivative of a function using the central difference as follows:

$$f'(x) \sim \frac{\Delta_c f(x)}{2\Delta x},$$

or in terms of the sampled sequence

$$(9.25) \qquad y(k) = \frac{f(k+1) - f(k-1)}{2\Delta x}.$$

Using Equations (9.1) and (9.4) with $L = 1$, $a_{-1} = -1/2\Delta x$, $a_0 = 0$, and $a_1 = 1/2\Delta x$, we determine the transfer function to be

$$H(w) = \frac{e^{2\pi i w \Delta x} - e^{-2\pi i w \Delta x}}{2\Delta x},$$

or

$$(9.26) \qquad H(w) = i\frac{\sin 2\pi w\Delta x}{\Delta x} = 2\pi i w \operatorname{sinc} 2\pi w\Delta x.$$

In Figure 9.9 we have plotted this transfer function (solid line) along with the true transfer function (dashed line), both normalized by Δx.

As can be seen from the figure, this algorithm fails rather miserably at the higher frequencies. However, there may be times when we wish to suppress the higher frequencies (e.g., a noisy signal) and then this would indeed be the desired algorithm. A measure of the accuracy of this algorithm is obtained by dividing Equation (9.26) by the true transfer function [Equation (9.15)]. This results in

$$(9.27) \qquad\qquad\qquad \operatorname{sinc} 2\pi w\Delta x,$$

which is plotted in Figure 9.10.

The second central difference of a function is defined as

$$\Delta_c^2 f(x) = \Delta_c[\Delta_c f(x)] = \Delta_c[f(x+\Delta x) - f(x-\Delta x)]$$

$$= f(x+2\Delta x) - 2f(x) + f(x-2\Delta x).$$

This second central difference is used to approximate the second difference of a function as follows:

$$y(k) = \frac{f(k+2) - 2f(k) + f(k-2)}{4\Delta x^2}.$$

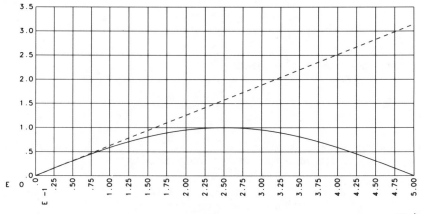

FIGURE 9.9. First central difference transfer function.

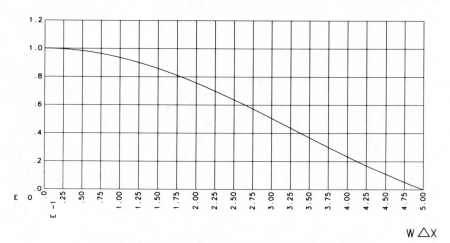

FIGURE 9.10. Accuracy of first central difference algorithm.

As we have done so many times before, we use Equations (9.1) and (9.4) to derive the transfer function of the preceding equation:

$$(9.28) \qquad H(w) = (i)^2 \left(\frac{\sin 2\pi w \Delta x}{\Delta x} \right)^2 = (2\pi i w)^2 (\text{sinc } 2\pi w \Delta x)^2$$

This transfer function is shown in Figure 9.11 (solid line) along with the actual transfer function of the second derivative (dashed line), both normalized

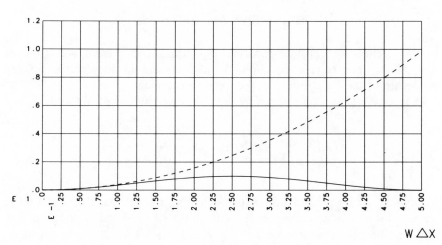

FIGURE 9.11. Second central difference transfer function.

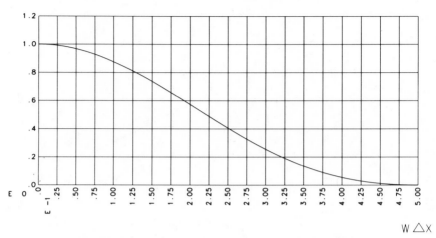

W △X

FIGURE 9.12.   Accuracy of second central difference formula.

by $(\Delta x)^2$. The normalized accuracy formula for this transfer function is given by

$$\left(\text{sinc } 2\pi w\Delta x\right)^2,$$

which is plotted in Figure 9.12.

In this section we have shown how to determine the transfer function of several derivative algorithms. Other algorithms would be handled in an analogous way.

RECURSIVE ALGORITHMS

In this section we consider recursive algorithms and present a method of determining their transfer function. We also discuss linearity and invariance properties of these algorithms in the frequency domain.

A recursive algorithm is given by the following mathematical formula:

$$y(k) = \sum_{l=-L}^{L} a_l E^l f(k) - \sum_{\substack{m=-M \\ m\neq 0}}^{M} b_m E^m y(k).$$

Examining the right-hand side of this equation, we see that the first term is simply that of a nonrecursive system (Equation 9.1) and the second term is a linear combination of the output (or feedback) terms. A slight algebraic manipulation of this equation yields the following, more useful, form:

$$(9.29) \qquad \sum_{m=-M}^{M} b_m E^m y(k) = \sum_{l=-L}^{L} a_l E^l f(k).$$

We now wish to determine the transfer function of this system. That is, we want to find $H(w)$ such that when $f(k) = \exp(2\pi i w k \Delta x)$, then $y(k) = H(w)\exp(2\pi i w k \Delta x)$. We begin by substituting

$$f(k) = e(k) = e^{2\pi i w k \Delta x}$$

into Equation (9.29). This results in

$$(9.30) \qquad \sum_{m=-M}^{M} b_m E^m y(k) = A(w)e(k),$$

where $A(w)$ is simply the transfer function of the nonrecursive portion of the system as per Equation (9.4):

$$A(w) = \sum_{l=-L}^{L} a_l e^{2\pi i w l \Delta x}.$$

Now let $y(k) = H(w)e(k) = H(w)\exp(2\pi i w k \Delta x)$, and substitute this into Equation (9.30) which gives us

$$\left(\sum_{m=-M}^{M} b_m e^{2\pi i w m \Delta x} \right) H(w)e(k) = A(w)e(k),$$

or

$$(9.31) \qquad G(w)H(w)e(k) = A(w)e(k),$$

where

$$(9.32) \qquad G(w) = \sum_{m=-M}^{M} b_m e^{2\pi i w m \Delta x}.$$

Solving for the transfer function $H(w)$ in Equation (9.31) we find

$$(9.33) \qquad H(w) = \frac{A(w)}{G(w)}.$$

Thus the transfer function of a recursive algorithm is given by the quotient of Equation (9.33), where $A(w)$ is the transfer function of the nonrecursive portion of the algorithm [Equation (9.4)] and $G(w)$ is the transfer function of the "feedback" portion of the system (Equation 9.32). As an example, let us consider the specific system described by

$$y(k + 1) = f(k) + f(k + 2) + y(k - 1),$$

or

$$(E^1 - E^{-1})y(k) = (1 + E^2)f(k).$$

Obviously, $A(w) = 1 + e^{4\pi i w \Delta x}$, and $G(w) = e^{2\pi i w \Delta x} - e^{-2\pi i w \Delta x}$. Therefore,

$$H(w) = \frac{A(w)}{G(w)} = \frac{1 + e^{4\pi i w \Delta x}}{e^{2\pi i w \Delta x} - e^{-2\pi i w \Delta x}}.$$

In the frequency domain we are now able to write

(9.34) $$Y(w) = \frac{A(w)}{G(w)} F(w).$$

Let us first look at the linearity of this equation. Consider $y_1(x)$ to be the output due to an input $f_1(x)$ and, similarly, $y_2(x)$ to be the resulting output when $f_2(x)$ is used as input. Also let their Fourier transforms be denoted as $Y_1(w)$, $F_1(w)$, $Y_2(w)$, and $F_2(w)$, respectively. Thus an input $f(x) = af_1(x) + bf_2(x)$ in the time domain becomes $F(w) = aF_1(w) + bF_2(w)$ in the frequency domain (theorem 3.1). When substituted into Equation (9.34) we obtain

$$Y(w) = \frac{A(w)}{G(w)} [aF_1(w) + bF_2(w)]$$

$$= a\frac{A(w)}{G(w)} F_1(w) + b\frac{A(w)}{G(w)} F_2(w) = aY_1(w) + bY_2(w).$$

In the time domain this becomes $y(x) = ay_1(x) + by(x)$. Thus we have demonstrated that a recursive system is linear with respect to the transfer function $A(w)/G(w)$.

Invariance with respect to this transfer function is demonstrated in an analogous manner. Let us delay the input function (or distribution) by an amount τ. That is, $\hat{f}(x) = f(x - \tau)$, which in the frequency domain is $\hat{F}(w) = F(w)\exp(-2\pi i w\tau)$. When substituted into Equation (9.34) we obtain

$$\hat{Y}(w) = \frac{A(w)}{G(w)} \hat{F}(w) = \frac{A(w)}{G(w)} F(w)e^{-2\pi i w\tau} = Y(w)e^{-2\pi i w\tau},$$

or in the time domain

$$\hat{y}(x) = y(x - \tau).$$

FINITE DIFFERENCE CALCULUS REVIEW

In this section we make a brief detour and present a few very basic results from the field of finite difference calculus. These results enable us to gain valuable insight into the operation of both integration and interpolation algorithms discussed in the following sections.

We have already learned about the forward differences of a function and how they transform the function. By way of examples we demonstrated their effect on such well-known functions as $\sin x$ and $\ln(x)$. We now introduce a new type of function, known as *factorial forms*, and study their behavior under the forward differences.

We define the *factorial form of degree n* as

$$(9.35) \qquad x^{(n)} = x(x - 1)(x - 2) \cdots (x - n + 1).$$

A few examples follow:

$$x^{(2)} = x(x - 1) = x^2 - x.$$

$$x^{(3)} = x(x - 1)(x - 2) = x^3 - 3x^2 + 2x.$$

As is obvious from this definition, the factorial form of degree n is simply an nth degree polynomial whose roots are $0, 1, \ldots, n - 1$.

We now consider the forward difference (with $\Delta x = 1$) of this new function. First for the case when $n = 2$,

$$\Delta x^{(2)} = (x + 1)^{(2)} - x^{(2)}, = (x + 1)x - x(x - 1)$$

$$= [(x + 1) - (x - 1)]x = 2x^{(1)}.$$

For $n = 3$,

$$\Delta x^{(3)} = (x + 1)^{(3)} - x^{(3)},$$

$$= (x + 1)(x)(x - 1) - x(x - 1)(x - 2),$$

$$= [(x + 1) - (x - 2)]x(x - 1),$$

$$\Delta x^{(3)} = 3x^{(2)}.$$

In general, we have

$$(9.36) \qquad \Delta x^{(n)} = nx^{(n-1)}.$$

This is easily verified as follows:

$$\Delta x^{(n)} = (x + 1)^{(n)} - x^{(n)},$$

$$= (x + 1)x(x - 1) \cdots (x - n) - x(x - 1) \cdots (x - n + 1),$$

$$= [(x + 1) - (x - n + 1)]x(x - 1)(x - 2) \cdots (x - n),$$

$$\Delta x^{(n)} = nx^{(n-1)}.$$

Higher differences of the factorial form result in

$$\Delta^2 x^{(n)} = n(n - 1)x^{(n-2)},$$

$$\Delta^3 x^{(n)} = n(n - 1)(n - 2)x^{(n-3)},$$

$$\vdots$$

$$\Delta^n x^{(n)} = n!.$$

It is hard not to notice the similarity between the forward difference of $x^{(n)}$ and the derivative of x^n. That is, $x^{(n)}$ plays an analogous role in the calculus of finite differences (with respect to the forward difference) to that of x^n in the infinitesimal calculus (with respect to the derivative).

Theorem 9.1 (Newton). If $f(x)$ is an nth degree polynomial in x, then it may be written as

$$f(x) = f(0) + x^{(1)}\Delta f(0) + \frac{x^{(2)}}{2!}\Delta^2 f(0) + \cdots + \frac{x^{(n)}}{n!}\Delta^n f(0),$$

where $\Delta^i f(0)$ is the ith forward difference of the function with $\Delta x = 1$.

Proof. By assumption, $f(x) = b_0 + b_1 x + \cdots + b_n x^n$, but since $x^{(i)}$ is a polynomial of degree i, we can always determine constants $a_0, \ldots, a_n$ such that

$$f(x) = a_0 + a_1 x^{(1)} + a_2 x^{(2)} + \cdots + a_n x^{(n)}.$$

Now differencing $f(x)$ n times (with $\Delta x = 1$) we obtain

$$\Delta f(x) = a_1 + 2a_2 x^{(1)} + 3a_3 x^{(2)} + \cdots + na_n x^{(n-1)}$$

$$\Delta^2 f(x) = 2a_2 + 3(2)a_3 x^{(1)} + \cdots + n(n - 1)a_n x^{(n-2)}$$

$$\vdots$$

$$\Delta^n f(x) = n!a_n.$$

If we now let $x = 0$ in this equation we obtain

$$a_0 = f(0),$$

$$a_1 = \Delta f(0),$$

$$a_2 = \frac{\Delta^2 f(0)}{2!},$$

$$\vdots$$

$$a_n = \frac{\Delta^n f(0)}{n!}.$$

As an example, suppose that the data in Table 9.1 was generated using a third degree polynomial and that we wish to find an expression for that polynomial. We proceed as follows:

$$f(x) = f(0) + \Delta f(0) x^{(1)} + \frac{\Delta^2 f(0)}{2!} x^{(2)} + \frac{\Delta^3 f(0)}{3!} x^{(3)}.$$

$$f(0) = 1,$$

$$\Delta f(0) = f(1) - f(0) = 3,$$

$$\Delta^2 f(0) = f(2) - 2f(1) + f(0) = 3,$$

$$\Delta^3 f(0) = f(3) - 3f(2) + 3f(1) - f(0) = 1.$$

Thus

$$f(x) = 1 + 3x + \tfrac{3}{2}x(x - 1) + \tfrac{1}{6}x(x - 1)(x - 2),$$

$$f(x) = \tfrac{1}{6}[x^3 + 6x^2 + 11x + 6].$$

TABLE 9.1. Newton's Theorem Example

x	$f(x)$
0	1
1	4
2	10
3	20
4	35
5	56

As our next example, let us express the polynomial

$$f(x) = 2x^3 - 3x^2 + 3x - 10$$

as a series of factorial forms. To do so we proceed as follows:

$$f(x) = f(0) + \Delta f(0)x^{(1)} + \frac{\Delta^2 f(0)}{2!}x^{(2)} + \frac{\Delta^3 f(0)}{3!}x^{(3)},$$

$$f(0) = -10,$$

$$\Delta f(0) = f(1) - f(0) = 2,$$

$$\Delta^2 f(0) = f(2) - 2f(1) + f(0) = 12,$$

$$\Delta^3 f(0) = f(3) - 3f(2) + 3f(1) - f(0) = 12.$$

Thus

$$f(x) = 2x^{(3)} + 6x^{(2)} + 2x^{(1)} - 10.$$

INTEGRATION ALGORITHMS

In this section we use Newton's theorem to derive a few integration algorithms and then study their transfer functions.

Shown in Figure 9.13 is a function $f(x)$ whose domain has been divided into intervals of equal width Δx. The problem at hand is to determine the area "under" the curve described by $f(x)$. Mathematically, the area from $k\Delta x$ to $(k + n)\Delta x$ is given as

$$A_n = \int_{k\Delta x}^{(k+n)\Delta x} f(x)\, dx.$$

We now make a change of variable and let

$$t = \frac{x - k\Delta x}{\Delta x},$$

or $dx = \Delta x\, dt$. Note also that when $x = k\Delta x$ then $t = 0$ and when $x = (k + n)\Delta x$ then $t = n$. Therefore, the preceding integral becomes

$$A_n = \Delta x \int_0^n f(t)\, dt.$$

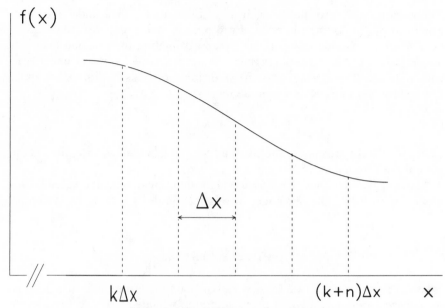

FIGURE 9.13. Function to be integrated.

The next step is to use Newton's theorem and approximate this function (from $t = 0$ to $t = n$) by an nth degree polynomial:

$$f(t) = f(0) + \Delta f(0)t^{(1)} + \cdots + \frac{\Delta^n f(0)}{n!}t^{(n)}.$$

When this is substituted into the integral equation and the actual integration is performed we obtain

$$A_n = \Delta x \left[nf(0) + \frac{n^2}{2}\Delta f(0) + \left(\frac{n^3}{3} - \frac{n^2}{2} \right)\frac{\Delta^2 f(0)}{2!} \right.$$
$$\left. + \left(\frac{n^4}{4} - n^3 + n^2 \right)\frac{\Delta^3 f(0)}{3!} + \cdots \right]$$

or, noting the fact that for $t = 0$ we have $x = k\Delta x$, we obtain

$$(9.37) \qquad A_n = \Delta x \left[nf(k) + \frac{n^2}{2}\Delta f(k) + \left(\frac{n^3}{3} - \frac{n^2}{2} \right)\frac{\Delta^2 f(k)}{2!} \right.$$
$$\left. + \left(\frac{n^4}{4} - n^3 + n^2 \right)\frac{\Delta^3 f(k)}{3!} + \cdots \right]$$

where (as always) $f(k)$ means the function evaluated at $x = k\Delta x$.

Equation (9.37) provides us with an estimate of the area under the curve generated by $f(x)$ over an interval of width $n\Delta x$; that is, the area from $k\Delta x$ to $(k + n)\Delta x$. It obtains this estimate by approximating the function over this interval with an nth degree polynomial. To determine the total area under $f(x)$ we simply divide its domain into several disjoint intervals of width $n\Delta x$, apply Equation (9.37) to each, and sum the results:

$$(9.38) \qquad\qquad y(k + n) = y(k) + A_k,$$

where $y(k)$ is the area under the curve up to, and including, $k\Delta x$ and A_k is the area of $f(x)$ from $k\Delta x$ to $(k + n)\Delta x$.

Many of the well-known integration algorithms are obtained as special cases of Equation (9.37). The remainder of this section deals with such algorithms.

The Trapezoidal Rule

When in Equation (9.37) we set $n = 1$, we obtain the trapezoidal rule. That is, we choose an n-interval width of Δx and approximate $f(x)$ by a first degree polynomial (i.e., a straight line) over this interval. Thus in Equation (9.37) we let $n = 1$ and ignore all differences above $\Delta f(k)$. This results in

$$A = \Delta x\left\{ f(k) + \frac{1}{2}[f(k + 1) - f(k)]\right\} = \frac{\Delta x}{2}[f(k + 1) + f(k)].$$

When this is substituted into Equation (9.38) we obtain the well-known trapezoidal rule:

$$y(k + 1) = y(k) + \frac{\Delta x}{2}[f(k + 1) + f(k)],$$

or, in the form of a recursive algorithm (Equation 9.29),

$$(E^1 - 1)y(k) = \frac{\Delta x}{2}(E^1 + 1)f(k).$$

Using Equation (9.33) we can easily write the transfer function of this algorithm as

$$H(w) = \frac{\Delta x}{2}\frac{e^{2\pi i w\Delta x} + 1}{e^{2\pi i w\Delta x} - 1},$$

or, with a little complex variable algebra,

$$(9.39) \qquad\qquad H(w) = \frac{\Delta x \cos \pi w\Delta x}{2i \sin \pi w\Delta x}.$$

As usual, it is instructive to compare this result with the true answer and thus we must determine the transfer function of the integral operation. Let us assume an input of $e(x) = \exp[2\pi i w x]$ and calculate the output $y(x)$ for an integrating system:

$$y(x) = \int e^{2\pi i w t}\, dt.$$

Ignoring initial conditions, we can integrate this equation to obtain

$$y(x) = \frac{e^{2\pi i w x}}{2\pi i w} = \frac{1}{2\pi i w} e(x).$$

Therefore, we see

(9.40) $$H(w) = \frac{1}{2\pi i w}.$$

Shown in Figure 9.14 is a plot of both the trapezoidal integration algorithm transfer function of Equation (9.39) (solid line) along with the true transfer function of Equation (9.40). This plot provides a picture of the accuracy of the algorithm. A better measure of the accuracy can be obtained by dividing Equation (9.39) by Equation (9.40) which results in the normalized form

(9.41) $$\frac{\cos \pi w \Delta x}{\text{sinc } \pi w \Delta x}.$$

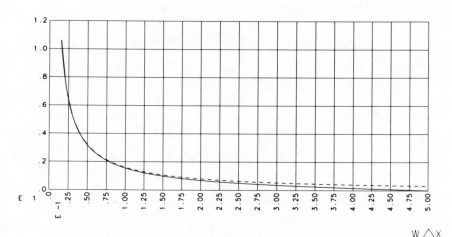

W $\triangle$ X

FIGURE 9.14. Trapezoidal rule transfer function.

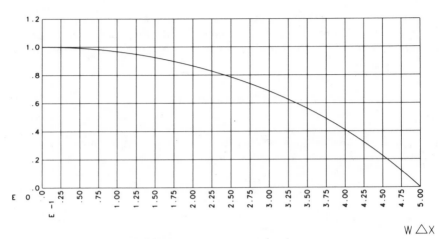

FIGURE 9.15. Accuracy of trapezoidal rule.

Shown in Figure 9.15 is a plot of this equation which clearly shows that the trapezoidal rule has the effect of suppressing the higher frequencies.

Simpson's $\frac{1}{3}$ Rule

If now in Equation (9.37) we set $n = 2$, we obtain Simpson's $\frac{1}{3}$ rule. Physically, we choose an n-interval of width $2\Delta x$ and approximate $f(x)$ by a second degree polynomial over this interval. Therefore, in Equation (9.37) when we set $n = 2$ and ignore differences above $\Delta^2 f(k)$ we obtain

$$A = \Delta x\{2f(k) + 2[f(k + 1) - f(k)]$$

$$+ \tfrac{2}{3}[f(k + 2) - 2f(k + 1) + f(k)]\},$$

or, after a little simplifying,

$$A = \frac{\Delta x}{3}[f(k + 2) + 4f(k + 1) + f(k)].$$

Substituting this result into Equation (9.38) we obtain our desired result:

$$y(k + 2) - y(k) = \frac{\Delta x}{3}[f(k + 2) + 4f(k + 1) + f(k)],$$

or, in recursive algorithm form,

$$(E^2 - 1)y(k) = \frac{\Delta x}{3}(E^2 + 4E + 1)f(k).$$

This results in the transfer function

$$H(w) = \frac{\Delta x}{3} \frac{e^{4\pi i w \Delta x} + 4e^{2\pi i w \Delta x} + 1}{e^{4\pi i w \Delta x} - 1}.$$

After simplifying this becomes

(9.42) $$H(w) = \frac{\Delta x}{i} \frac{2 + \cos 2\pi w \Delta x}{3 \sin 2\pi w \Delta x}.$$

In Figure 9.16 we have plotted this transfer function (solid line) and the true transfer function (dashed line). When we divide Equation (9.42) by Equation (9.40) we obtain

(9.43) $$\frac{2 + \cos 2\pi w \Delta x}{3 \text{sinc} \, 2\pi w \Delta x},$$

which is our measure of accuracy for Simpson's $\frac{1}{3}$ Rule. It is shown in Figure 9.17, where we notice the unexpected result, that this algorithm has the effect of drastically enhancing the higher frequencies.

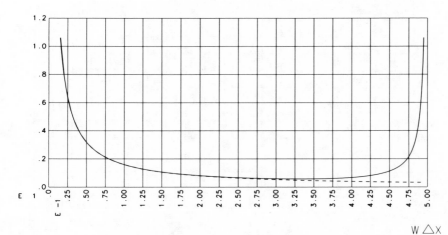

FIGURE 9.16. Transfer function of Simpson's $\frac{1}{3}$ rule.

Simpson's $\frac{3}{8}$ Rule

When we choose an *n*-interval of width $3\Delta x$ and approximate $f(x)$ by a third degree polynomial, we obtain Simpson's $\frac{3}{8}$ rule. That is, we set $n = 3$ in Equation (9.37) and ignore all differences above $\Delta^3 f(k)$:

$$A = \Delta x\{3f(k) + \tfrac{9}{2}[f(k + 1) - f(k)] + \tfrac{9}{4}[f(k + 2) - 2f(k + 1) + f(k)]$$

$$+ \tfrac{3}{8}[f(k + 3) - 3f(k + 2) + 3f(k + 1) - f(k)]\}.$$

After rearrangement we have

$$A = \frac{3\Delta x}{8}[f(k + 3) + 3f(k + 2) + 3f(k + 1) + f(k)]$$

which when substituted into Equation (9.38) results in

$$y(k + 3) - y(k) = \frac{3\Delta x}{8}[f(k + 3) + 3f(k + 2) + 3f(k + 1) + f(k)].$$

Going through the same moves as before, we determine the transfer function of this algorithm to be

(9.44) $$H(w) = \frac{3\Delta x}{8i}\,\frac{\cos 3\pi w\Delta x + 3\cos \pi w\Delta x}{\sin 3\pi w\Delta x}.$$

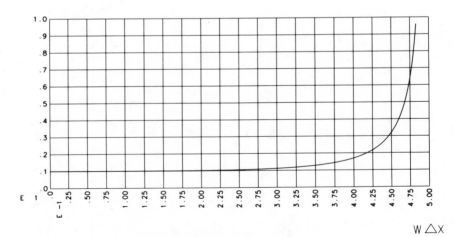

FIGURE 9.17. Accuracy of Simpson's $\frac{1}{3}$ rule.

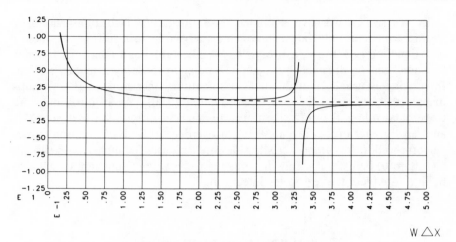

FIGURE 9.18. Transfer function of Simpson's $\frac{3}{8}$ rule.

This transfer function is shown in Figure 9.18 (solid line) along with the actual transfer function (dashed line). The accuracy formula for this algorithm is

$$(9.45) \qquad \frac{\cos 3\pi w\Delta x + 3\cos \pi w\Delta x}{4 \operatorname{sinc} 3\pi w\Delta x}$$

which is shown in Figure 9.19. As can be seen from these figures, this algorithm has its problems at $w = \Delta x/3$, where it blows up.

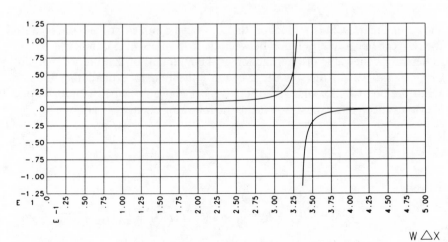

FIGURE 9.19. Accuracy of Simpson's $\frac{3}{8}$ rule.

In this section we have examined three integration algorithms and their transfer functions. Other algorithms can be derived by using different values of n in Equation (9.37) (e.g., $n = 6$ yields Weddle's rule) and then their transfer functions can be obtained by methods similar to those used in this section. Integration is usually thought of as a smoothing operation. That is, it suppresses the higher frequencies [see Equation (9.40)]. As it turns out, the trapezoidal rule is the only algorithm presented in this section that does this and, as a matter of fact (Figure 9.15), it overdoes it. Both of Simpson's rules perform well at low frequencies but deviate seriously for higher ones. Which algorithm to be used certainly depends upon the type of function, or data, to be integrated and the reason for integrating it.

INTERPOLATION ALGORITHMS

In this section we once again use Newton's theorem to derive a few interpolation algorithms and then look at their behavior in the frequency domain. Let us assume that we have a function that has been sampled with sampling rate $2\Delta x$. That is, we have values of this function only at discrete locations $x = k2\Delta x$. The purpose of interpolating a function is to estimate values of the function within these sampled intervals. Although we often use sophisticated mathematical techniques to determine these values, it nevertheless must be clearly understood that they are only our best guess.

Linear Interpolation

Shown in Figure 9.20 is a function that has been sampled at the values $k2\Delta x$. We now use linear interpolation between the two values $k2\Delta x$ and $(k + 1)2\Delta x$ to determine the value of the function at the midpoint of the interval, namely, $k2\Delta x + \Delta x$. To do this we use Newton's theorem to expand the function about the point $k2\Delta x$, and set all differences above Δf equal to zero. To use Newton's theorem we must first make the following change of variable:

$$(9.46) \qquad t = \frac{x - k2\Delta x}{2\Delta x}.$$

Thus

$$y(t) = f(0) + t\Delta f(0) = (1 - t)f(0) + tf(1),$$

where $y(t)$ is the interpolated value of the function at any location within the

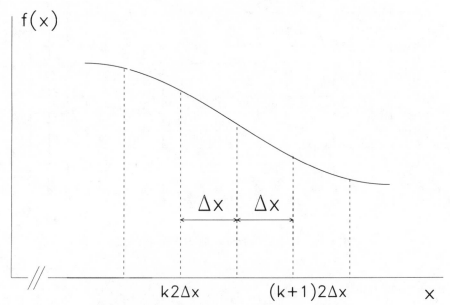

FIGURE 9.20.   Sampled function to be linearly interpolated.

interval. Now, of interest to us is the case when $t = \frac{1}{2}$ which yields the equation

$$y(\tfrac{1}{2}) = \frac{f(0) + f(1)}{2}.$$

We convert back to the unnormalized variable x by noting that in Equation (9.46) if we let $t = 0$, $\frac{1}{2}$, and 1, then we will obtain values of $x = 2k\Delta x$, $(2k + 1)\Delta x$, and $(2k + 2)\Delta x$, respectively. Finally, if we let $j = 2k$, our linear interpolation algorithm becomes

$$(9.47) \qquad y(j + 1) = \frac{f(j + 2) + f(j)}{2}, \qquad j \text{ is an even integer.}$$

A few comments are in order at this point. We note that the original sampled function has a sampling rate, or interval size, of $2\Delta x$ which corresponds to a maximum possible frequency of $1/4\Delta x$. The effect of the interpolation algorithm is to add "artificial" data points midway between the original ones. Thus our interpolated function has an interval size of Δx or, conversely, a maximum possible frequency of $1/2\Delta x$. Therefore, any frequencies between $1/4\Delta x$ and $1/2\Delta x$ have been added by the interpolation algorithm. To gain a better appreciation of this we can use our standard techniques and determine

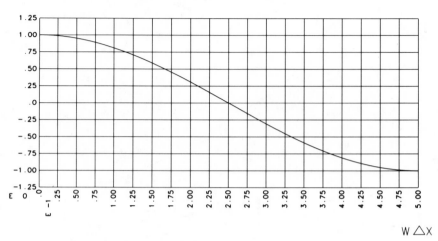

W $\triangle$X

FIGURE 9.21.   Transfer function of linear interpolation algorithm.

the transfer function to be

$$(9.48) \qquad\qquad H(w) = \cos 2\pi w\Delta x.$$

This is shown in Figure 9.21. As can be appreciated from this figure, the linear interpolation algorithm passes both the low frequencies and the high (artificial) frequencies. However, the midrange frequencies are rather severely attenuated. If the original function contained any high frequency "noise," then this linear interpolation algorithm will enhance it. However, if the original function was sampled properly, then there should be no frequencies above $1/4\Delta x$ present and, consequently, we need only concern ourselves with frequencies up to $w\Delta x = 0.25$.

Cubic Interpolation

In our previous or linear scheme we interpolated to the value of the function at the midpoint of the original sampling interval by passing a straight line between the two adjacent endpoints. For cubic interpolation we pass a cubic equation, or third degree polynomial, through the *two* adjacent interval endpoints on either side (see Figure 9.22). Again we use Newton's theorem to expand the function about $k\,2\Delta x$ and set all differences above $\Delta^3 f$ equal to 0. Making the change of variable dictated by Equation (9.46) we obtain

$$y(t) = f(0) + t^{(1)}\Delta f(0) + t^{(2)}\,\Delta^2 f(0) + t^{(3)}\,\Delta^3 f(0).$$

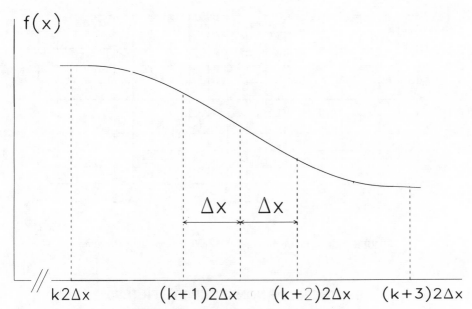

FIGURE 9.22. Cubic interpolation of a function.

This equation gives the interpolated value $y(t)$ of the function at any point t in the interval [0, 3] (or any point x in the interval $[k\,2\Delta x, (k + 3)2\,\Delta x]$). When we set $t = \frac{3}{2}$ and perform a little algebra we find

$$y(\tfrac{3}{2}) = \tfrac{1}{16}[-f(3) + 9f(2) + 9f(1) - f(0)].$$

Back in the x domain (Equation 9.46), this becomes the cubic interpolation algorithm

$$(9.49) \quad y(j + 3) = \tfrac{1}{16}[-f(j + 6) + 9f(j + 4) + 9f(j + 2) - f(j)].$$

In the frequency domain this algorithm has a transfer function given by

$$(9.50) \qquad H(w) = \frac{9\cos 2\pi w \Delta x - \cos 6 \pi w \Delta x}{8},$$

shown in Figure 9.23.

This algorithm also has the effect of passing both the low and high frequencies while suppressing the midrange ones. Other algorithms can, and should, be studied (in the frequency domain) to see how they interpolate the missing data.

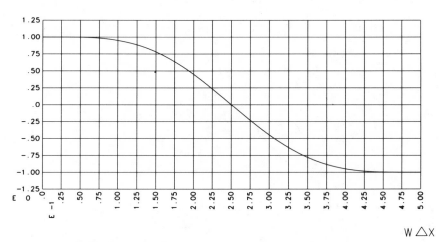

FIGURE 9.23. Transfer function of the cubic interpolation algorithm.

SYNTHESIS OF NONRECURSIVE FILTERS

Up to this point in the chapter, our approach has been to present numerical analysis algorithms in the "time domain" and then derive their transfer functions which we examine in the frequency domain. In this section we reverse the procedure and begin with a desired transfer function and then determine the algorithm to which it belongs.

Let us assume that we desire a particular transfer function, say $\hat{H}(w)$ over the frequency domain $[-\Omega, \Omega]$. The task before us is to design an algorithm that will exactly generate this transfer function. In general, the task is hopeless. However, we can often design an algorithm that has a transfer function that will very closely approximate our desired result. The approach that we take is to protract the transfer function $\hat{H}(w)$ to be periodic (see Chapter 2) and then represent it with a Fourier series over the protracted interval $[-\Omega, \Omega]$:

$$\hat{H}(w) = \sum_{l=-\infty}^{\infty} C_l e^{2\pi i w l/2\Omega},$$

where

$$C_l = \frac{1}{2\Omega} \int_{-\Omega}^{\Omega} \hat{H}(w) e^{-2\pi i w l/2\Omega} \, dw, \qquad l \in (-\infty, \infty).$$

For reasons that will soon become obvious, we only use a truncated Fourier series and *approximate* $\hat{H}(w)$ over the interval $[-\Omega, \Omega]$. Mathematically, that is, we approximate $\hat{H}(w)$ by $H(w)$ where

$$H(w) = \sum_{l=-L}^{L} C_l e^{2\pi i w l/2\Omega},$$

and

$$C_l = \frac{1}{2\Omega} \int_{-\Omega}^{\Omega} H(w) e^{-2\pi i w l/2\Omega} \, dw, \qquad l \in [-L, L].$$

We now assume that our algorithm is to operate on a function that has been sampled in accordance with the sampling Theorem 5.1 (i.e., $\Delta x = 1/2\Omega$). Thus our equations become

$$(9.51) \qquad H(w) = \sum_{l=-L}^{L} C_l e^{2\pi i w l \Delta x},$$

where

$$(9.52) \qquad C_l = \Delta x \int_{-1/2\,\Delta x}^{1/2\,\Delta x} H(w) e^{-2\pi i w \Delta x} \, dw \qquad l \in [-L, L].$$

If we now run this "backwards" through Equations (9.4) and (9.1), we obtain our algorithm:

$$(9.53) \qquad y(k) = \sum_{l=-L}^{L} C_l E^l f(k) = \sum_{l=-L}^{L} C_l f(k + l).$$

Thus we see the Fourier series components of the transfer function $H(w)$ are the coefficients of our nonrecursive filter.

Generally, we consider a nonrecursive filter to be an algorithm that selectively modifies a particular set of frequencies. The smoothing algorithms that we previously discussed tended to pass the low frequency components of a function while rejecting, or at least attenuating, the high ones. The ideal low pass filter is the pulse function

$$H(w) = \begin{cases} 1, & |w| \leqslant w_0 \\ 0 & \text{otherwise.} \end{cases}$$

This one will pass perfectly all the frequencies up to and including w_0, whereas those above this value are completely removed. Using Equation (9.52), or simply recalling previous calculations from Chapter 2, we can obtain the

Fourier series coefficients

$$C_l = 2w_0 \Delta x \operatorname{sinc} 2\pi l w_0 \Delta x.$$

When these values are substituted into Equation (9.53), we obtain an algorithm that will approximate this low pass filter. The trouble with this algorithm is that it requires a large number of terms to give a reasonable approximation and even then the ringing, or Gibb's effect, extends well out into the higher frequencies. Shown in Figure 9.24 is the approximation to this filter for $L = 12$. *Note*: w_0 has been normalized to $\frac{1}{2}$ of Ω which, in turn, is normalized to unity.

Let us now consider another low pass filter given by the Gaussian formula

(9.54) $$H(w) = e^{-(w/w_0)^2}.$$

To determine the algorithm coefficients of this filter we could use Equation (9.52). However, we use instead some results from Chapter 5 and determine them directly from its Fourier transform which is given by (Equation 3.16)

$$w_0 \sqrt{\pi}\, e^{-\pi^2 \xi^2 w_0^2}.$$

Thus using Equation (5.12) we find

$$C_l = \frac{w_0 \sqrt{\pi}}{2\Omega} e^{-\pi^2 l^2 w_0^2 / 4\Omega^2},$$

or

$$C_l = \Delta x w_0 \sqrt{\pi}\, e^{-\pi^2 l^2 w_0^2 (\Delta x)^2}.$$

The advantage of this algorithm is that it requires relatively few terms to give a reasonable approximation to the Gaussian filter. The disadvantage of

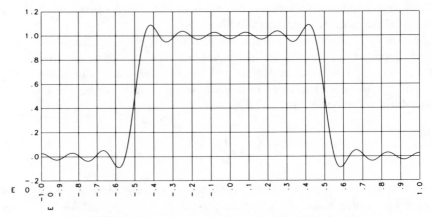

FIGURE 9.24.   24-term approximation to the pulse filter algorithm.

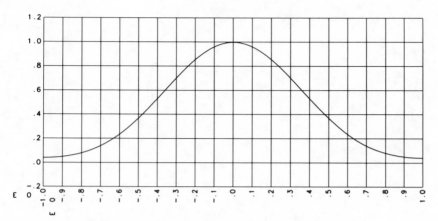

FIGURE 9.25. Seven-term approximation to the Gaussian filter.

this filter is that it does not have a sharp cutoff frequency w_0, as does the pulse function, but instead "rolls off" the frequencies gradually. Shown in Figure 9.25 is a seven-term ($L = 3$) approximation to the Gaussian filter. (*Note*: w_0 has been normalized to $\frac{1}{2}$ the value of Ω which, in turn, is normalized to 1).

SUMMARY

In this chapter we have shown how to consider various numerical analysis algorithms from a frequency domain point of view. We learned that much insight into an algorithm's behavior can be obtained by an examination of its transfer function. Also presented in this chapter was a very useful and powerful method of designing nonrecursive filters. Obviously, it is impossible to discuss all possible algorithms here but the material presented in this chapter should serve as a guide to the analysis of any other linear algorithm.

BIBLIOGRAPHY

Raven, F. H., *Mathematics of Engineering Systems*. McGraw-Hill, New York, 1966.

Richardson, C. H., *An Introduction to the Calculus of Finite Differences*. Van Nostrand, Princeton, N.J., 1954.

Ralston, A., *A First Course in Numerical Analysis*. McGraw-Hill, New York, 1965.

Hamming, R. W., *Digital Filters*. Prentice-Hall, Englewood Cliffs, N.J., 1977.

Oppenheim, A. V. and R. W. Schafer, *Digital Signal Processing*. Prentice-Hall, Englewood Cliffs, N.J., 1975.

Weaver, H. J., *Simple Nonrecursive Algorithms to Approximate Super-Gaussian Filters*. Report No. UCID-17962, University of California, Lawrence Livermore Laboratory, Livermore, Calif., November, 1978.

10

THE HEAT EQUATION

In this chapter we examine the heat, or diffusion, equation and illustrate how to solve several "typical" problems by applying Fourier analysis to this equation. Although this equation actually describes many diverse phenomena such as neutron transport in an homogeneous material, current and voltage transfer in ideal transmission lines, and vorticity diffusing through a fluid, we limit our discussion to applications in the area of heat transfer in an homogeneous body. The extension of this analysis to other fields is obtained by straightforward and simple analogous reasoning.

DERIVATION OF THE HEAT EQUATION

In this section we present a brief derivation of the heat equation in two dimensions. We begin with a portion of a solid homogeneous body, as illustrated in Figure 10.1, with a temperature distribution $u(x, y; t)$.

Inside this body we show a small element of width Δx and height Δy. The depth of the element will be denoted as l. The rate at which heat flows across the x-face of this element is described by the equation

$$q_x = -kl\Delta y \frac{\partial u(x, y; t)}{\partial x},$$

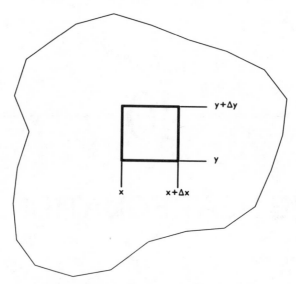

FIGURE 10.1. Portion of a homogeneous solid.

where k is the *thermal conductivity* of the material, Δyl is the surface area through which the heat flows, and $\partial u(x, y; t)/\partial x$ is the temperature gradient in the x-direction. Similarly, the rates of heat flow across the other surfaces are given by

$$q_{x+\Delta x} = -kl\Delta y \frac{\partial u(x + \Delta x, y; t)}{\partial x},$$

$$q_y = -kl\Delta x \frac{\partial u(x, y; t)}{\partial y},$$

$$q_{y+\Delta y} = -kl\Delta x \frac{\partial u(x, y + \Delta y; t)}{\partial y}.$$

The rate at which heat is accumulated within this element is simply the difference between the rate of heat flowing in and that flowing out. Mathematically, we have

$$\Delta q = q_x + q_y - q_{x+\Delta x} - q_{y+\Delta y},$$

(10.1) $$\Delta q = kl\Delta y \left[\frac{\partial u(x + \Delta x, y; t)}{\partial x} - \frac{\partial u(x, y; t)}{\partial x} \right]$$

$$+ kl\Delta x \left[\frac{\partial u(x, y + \Delta y; t)}{\partial y} - \frac{\partial u(x, y; t)}{\partial y} \right].$$

Now, using the *coefficient of heat capacity c*, we can also relate this rate of heat accumulation to the rate of change of the temperature with time. We do so by noting that a measure of the amount of heat contained within the volume $\Delta x \, \Delta yl$ is given by the well-known equation

$$Q = c\Delta x\Delta yl\, u\left(x + \frac{\Delta x}{2}, y + \frac{\Delta y}{2}; t\right).$$

Thus the rate of change of this contained (or accumulated) heat is obtained by differentiating the preceding equation with respect to time:

$$(10.2) \qquad \Delta q = \frac{\partial Q}{\partial t} = c\Delta x\Delta yl\frac{\partial u(x + \Delta x/2, \, y + \Delta y/2; \, t)}{\partial t}.$$

Equating Equations (10.1) and (10.2) and using a little algebra we find

$$c\frac{\partial u(x + \Delta x/2, \, y + \Delta y/2; \, t)}{\partial t} = \frac{k}{\Delta x}\left[\frac{\partial u(x + \Delta x, \, y; \, t)}{\partial x} - \frac{\partial u(x, \, y; \, t)}{\partial x}\right]$$

$$+ \frac{k}{\Delta y}\left[\frac{\partial u(x, \, y + \Delta y; \, t)}{\partial y} - \frac{\partial u(x, \, y; \, t)}{\partial y}\right].$$

Now, in the limit as both Δx and Δy approach zero, this equation becomes our desired result, namely, the heat equation in two dimensions:

$$(10.3) \qquad \frac{\partial u(x, \, y; \, t)}{\partial t} = \alpha\left(\frac{\partial^2 u(x, \, y; \, t)}{\partial x^2} + \frac{\partial^2 u(x, \, y; \, t)}{\partial y^2}\right), \quad \text{where } \alpha = \frac{k}{c}.$$

In the remainder of this chapter we deal with methods of solving this equation. Normally, which method to use will depend upon the particular boundary or initial constraints placed upon the problem. In the next section we derive the finite Fourier sine and finite Fourier cosine transforms which are extremely useful for problems in which the homogeneous medium has finite dimensions.

FINITE SINE AND COSINE TRANSFORMS

The finite Fourier sine and finite Fourier cosine transforms are in effect special cases of the Fourier series representation of a function. We proceed to first derive the sine transform. Let us assume that $f(x)$ is an odd function that is periodic over the interval $[-L, L]$. (Actually, in most problems we are only interested in the function from 0 to L). Referring to Chapter 2, we know

$(L = T/2)$ that the Fourier series representation of this function is given as

(10.4)
$$f(x) = \sum_{n=1}^{\infty} B_n \sin\frac{\pi n x}{L},$$

(10.5)
$$B_n = \frac{2}{L}\int_0^L f(x)\sin\frac{\pi n x}{L}\,dx.$$

We may consider Equations (10.4) and (10.5) to be transform pairs. That is, we may consider Equation (10.5) to define an operation that maps $f(x)$ to the sequence $\langle B_n\rangle$. Similarly, we consider Equation (10.4) to be the inverse mapping that takes the sequence $\langle B_n\rangle$ back to the original function $f(x)$. The operation, or transform, of Equation (10.5) is called the *finite Fourier sine transform* and is denoted as $F_s(n)$ or $S[f(x)]$. Keep in mind, however, that it is nothing more than the Fourier series representation of an odd function. Now, let us determine the finite Fourier sine transform of the second derivative of $f(x)$:

(10.6)
$$S[f^{[2]}(x)] = \frac{2}{L}\int_0^L f^{[2]}(x)\sin\frac{\pi n x}{L}\,dx.$$

Using integration by parts twice on this equation we can show

$$S[f^{[2]}(x)] = \frac{-n\pi}{L}\frac{2}{L}\left[f(x)\cos\frac{n\pi x}{L}\Big|_0^L + \frac{n\pi}{L}\int_0^L f(x)\sin\frac{n\pi x}{L}\,dx\right],$$

or

(10.7) $$S[f^{[2]}(x)] = -\frac{\pi n}{L}\left[\frac{\pi n}{L}F_s(n) + \frac{2}{L}(-1)^n f(L) - \frac{2}{L}f(0)\right].$$

Equation (10.7) tells us that we can express the finite sine transform of the second derivative of $f(x)$ in terms of the transform of $f(x)$ and *values* of the function at $x = L$ and $x = 0$. The usefulness of this transform lies in the fact that it facilitates the solution of a class of boundary value problems as we demonstrate in later sections.

We now develop the finite Fourier cosine transform. When $f(x)$ is an even periodic function over the interval $[-L, L]$, its Fourier series representation is given as

(10.8)
$$f(x) = A_0 + \sum_{n=1}^{\infty} A_n \cos\frac{\pi n x}{L},$$

(10.9)
$$A_0 = \frac{1}{L}\int_0^L f(x)\,dx,$$

$$A_n = \frac{2}{L}\int_0^L f(x)\cos\frac{\pi n x}{L}\,dx.$$

Equations (10.8) and (10.9) again represent a transform pair. More specifically, Equation (10.9) is called the *finite Fourier cosine transform* of $f(x)$ and is denoted as $F_C(n)$ or $C[f(x)]$. As with the finite sine transform, we are able to show

$$(10.10) \quad C[f^{[2]}(x)] = -\left[-\left(\frac{\pi n}{L}\right)^2 F_C(n) + \frac{2}{L}(-1)^n f^{[1]}(L) - \frac{2}{L}f^{[1]}(0)\right].$$

Thus Equation (10.10) tells us that the finite Fourier cosine transform of the second derivative of $f(x)$ can be expressed in terms of the transform of $f(x)$ and the *values* of the first derivative of $f(x)$ at $x = 0$ and $x = L$.

ONE-DIMENSIONAL HEAT FLOW IN FINITE BODIES

In this section we examine finite bodies in which the temperature distribution along one dimension (in our case y) is assumed to be constant and, therefore, $\partial^2 u(x, y; t)/\partial y^2 = 0$. A classical example of this type of problem is an extended thin fin. When this one-dimensional assumption is substituted into Equation (10.3) we obtain

$$(10.11) \quad \frac{\partial u(x; t)}{\partial t} = \alpha \frac{\partial^2 u(x; t)}{\partial x^2}.$$

Ends Held at Zero Temperature

As our first problem we study the temperature distribution in a finite body in which both ends are held at zero temperature (see Figure 10.2). We also assume that the initial temperature distribution is given by the function $f(x)$. Mathematically, these initial and boundary conditions may be stated as

$$u(x; 0) = f(x), \quad \text{where } f(0) = f(L) = 0,$$

$$u(0; t) = 0,$$

$$u(L; t) = 0.$$

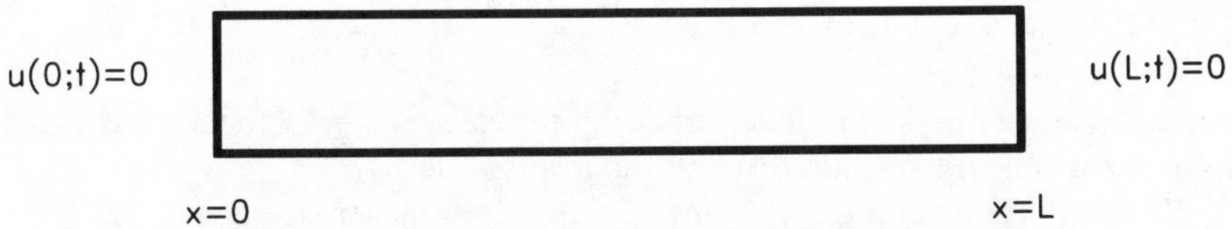

u(0;t)=0   u(L;t)=0

x=0   x=L

FIGURE 10.2. Finite body with ends held at zero temperature.

The approach that we take in solving this problem is to first take the finite Fourier sine transform of both sides of Equation (10.11) with respect to the variable x. We note here that when we transform a two-dimensional function with respect to only one of its variables we consider the other one to be held constant. (This is analogous to the line of reasoning that is used to calculate the partial derivative of a multidimensional function.) After transforming a two-dimensional function, the resulting expression is some new function of n and t that we denote as $U_s(n; t)$. To clarify these comments, our first example is discussed in some detail. We begin by multiplying both sides of Equation (10.11) by

$$\frac{2}{L} \sin \frac{\pi n x}{L}$$

and integrating the resulting expressions from 0 to L.

$$\frac{2}{L} \int_0^L \frac{\partial u(x; t)}{\partial t} \sin \frac{\pi n x}{L} \, dx = \alpha \frac{2}{L} \int_0^L \frac{\partial^2 u(x; t)}{\partial x^2} \sin \frac{\pi n x}{L} \, dx.$$

In this equation the integration of the functions $\partial u(x; t)/\partial t$ and $\partial^2 u(x; t)/\partial x^2$ is performed with respect to the variable x (t is held constant). It is also assumed that both of these functions satisfy the Dirichlet conditions (with respect to x) and, therefore, possess a finite transform. Referring to Equations (10.5) and (10.7) and using a little calculus, the equation becomes

$$\frac{\partial U_S(n; t)}{\partial t} = -\alpha \frac{\pi n}{L}\left[\frac{\pi n}{L} U_S(n; t) + \frac{2}{L}(-1)^n u(L; t) - \frac{2}{L} u(0; t)\right].$$

However, inasmuch as $u(L; t) = u(0; t) = 0$, we finally obtain

$$(10.12) \qquad \frac{\partial U_S(n; t)}{\partial t} + \alpha\left(\frac{\pi n}{L}\right)^2 U_S(n; t) = 0, \qquad n = 1, 2, \ldots.$$

Thus we have reduced our original problem to one of solving a set of (uncoupled) linear first-order differential equations. The solution to each of these equations is obtained by using classical methods and is given by

$$(10.13) \qquad U_S(n; t) = A_n e^{-\alpha(n\pi/L)^2 t}, \qquad n = 1, 2, \ldots,$$

where the constants A_n are determined by the initial ($t = 0$) constraints placed upon the problem (we soon discuss this in more detail).

We now have a solution in terms of the finite Fourier sine transform of the temperature distribution. To obtain the solution in terms of $u(x; t)$ itself, we

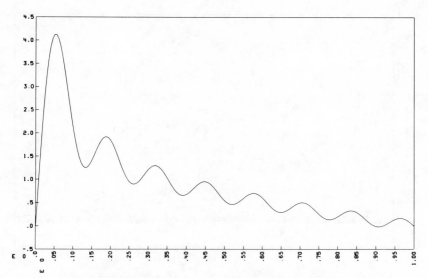

FIGURE 10.3.   Example of initial temperature profile.

simply use the (inverse transform) operation of Equation (10.4).

$$(10.14) \qquad u(x;t) = \sum_{n=1}^{\infty} A_n \sin \frac{\pi n x}{L} e^{-\alpha(\pi n/L)^2 t}.$$

If now in the preceding equation we let $t = 0$ (i.e., apply the initial conditions), we obtain

$$(10.15) \qquad u(x;0) = f(x) = \sum_{n=1}^{\infty} A_n \sin \frac{\pi n x}{L}.$$

The preceding equation tells us that the constants A_n may be considered the Fourier series coefficients of the initial temperature distribution function $f(x)$.[†] Therefore, they are described mathematically as

$$(10.16) \qquad A_n = \frac{2}{L} \int_0^L f(x) \sin \frac{\pi n x}{L} dx.$$

As an example, let us assume that the initial temperature distribution function

[†]Note that for our analysis we consider $f(x)$ to be an odd periodic function over the interval $[-L, L]$. Physically, however, we are only interested in its Fourier series over the half-interval $[0, L]$.

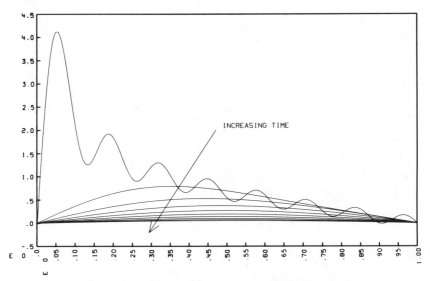

FIGURE 10.4. Example of temperature profile at various times.

$f(x)$ has a Fourier series frequency content described analytically as

$$A_n = \frac{1}{\sqrt{n}}$$

This function is shown in Figure 10.3. If we now substitute these values into Equation (10.14), we can describe the temperature profile of the body for all values of x and t. In Figure 10.4 we show this profile for several (equally spaced) increments in time. As can be seen from this figure [and appreciated from Equation (10.14)], the higher spatial temperature variations decay very rapidly.

Ends Held at Different Constant Temperatures

As our next problem we again consider the same finite body in Figure 10.2, only this time the end temperatures are not held at zero. More specifically, one end ($x = 0$) is held constant at temperature T_1 while the other end ($x = L$) remains fixed at T_2. Also, as in the previous example, we assume that the initial temperature profile is described by $f(x)$. Mathematically, these boundary conditions are given as

$$u(0; t) = T_1,$$

$$u(L; t) = T_2,$$

$$u(x; 0) = f(x), \qquad [f(0) = T_1, f(L) = T_2].$$

To solve this problem we again begin by taking the finite Fourier sine transform of Equation (10.11) which yields

$$\frac{dU_S(n;t)}{dt} = -\alpha\frac{\pi n}{L}\left[\frac{\pi n}{L}U_S(n;t) + \frac{2}{L}(-1)^n u(L;t) - \frac{2}{L}u(0;t)\right].$$

Applying the prevailing boundary conditions and after a little rearranging, we arrive at

$$\frac{dU_S(n;t)}{dt} + \alpha\left(\frac{\pi n}{L}\right)^2 U_S(n;t) = \frac{\alpha\pi n}{L}\left[T_1 - (-1)^n T_2\right], \qquad n = 1,2,\dots.$$

This time we have arrived at a set of (uncoupled) linear first-order general differential equations. The solution can again be obtained by classical methods[†] and is given by

$$(10.17) \quad U_S(n;t) = A_n e^{-\alpha(n\pi/L)^2 t} + \frac{2}{\pi n}\left[T_1 - (-1)^n T_2\right], \qquad n = 1,2,\dots.$$

where the constants A_n must be determined from the initial conditions.

Equation (10.17) gives us the answer in the transform domain and, therefore, to go back to the spatial domain we must use Equation (10.4).

$$(10.18) \quad u(x;t) = \sum_{n=1}^{\infty}\left\{A_n e^{-\alpha(\pi n/L)^2 t} + \frac{2}{\pi n}\left[T_1 - (-1)^n T_2\right]\right\}\sin\frac{\pi n x}{L}.$$

To determine the constants A_n we first let $t = 0$ (i.e., invoke the initial conditions) in the preceding equation. This move results in

$$u(x;0) = f(x) = \sum_{n=1}^{\infty}\left\{A_n + \frac{2}{\pi n}\left[T_1 - (-1)^n T_2\right]\right\}\sin\frac{\pi n x}{L}.$$

If now for the sake of simplifying the notation we let

$$C_n = A_n + \frac{2}{\pi n}\left[T_1 - (-1)^n T_2\right],$$

we obtain

$$f(x) = \sum_{n=1}^{\infty} C_n \sin\frac{\pi n x}{L}.$$

In this form we are able to easily recognize the constants C_n as the Fourier

[†]See the text by Raven listed in the Bibliography at the end of this chapter.

series coefficients of the initial temperature profile function $f(x)$:

$$C_n = \frac{2}{L} \int_0^L f(x) \sin \frac{\pi n x}{L} dx$$

or, in terms of the A_n constants,

(10.19) $A_n = \frac{2}{L} \int_0^L f(x) \sin \frac{\pi n x}{L} dx - \frac{2}{\pi n} \left[T_1 - (-1)^n T_2 \right].$

Together, Equations (10.18) and (10.19) completely describe the temporal and spatial thermal profile of a finite (one-dimensional) body whose ends are fixed at different constant temperatures. At first glance Equation (10.18) may not seem consistent with the boundary conditions because when $x = 0$ we see that $u(0; t) = 0$, and not T_1 as required. However, we should recall that to use the finite sine transform it was necessary to assume that the function $u(x; t)$ was an odd function (with respect to x) over the interval $[-L, L]$ (although we were only interested in the results from $[0, L]$). That is, we invented a ghost, or reflection, function from $-L$ to 0 and required that $u(-x; t) = -u(x; t)$. Certainly, when $u(0; t) \neq 0$ we must have a jump discontinuity at the location $x = 0$. As it turns out for a jump discontinuity at $x = a$ the Fourier series representation of a function converges to the average value of $f(a^-)$ and $f(a^+)$. In our case this average value is zero. Thus the value $u(0; t)$ must be calculated in the sense of a right-hand limit.

$$u(0; t) = \lim_{x \to 0^+} u(x; t).$$

This discontinuity at the origin can often be circumvented by making the

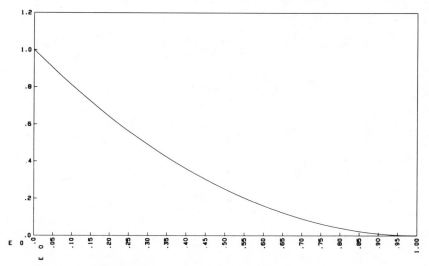

FIGURE 10.5.   Example of initial temperature profile.

change of variable

$$\hat{u}(x; t) = u(x; t) - T_1.$$

Similar remarks also hold for the situation in which $x = L$ and $u(L; t) = T_2 \neq 0$.

As an example, let us consider the initial temperature profile function described by (see Figure 10.5)

$$f(x) = T_1\left(\frac{L - x}{L}\right)^2, \qquad \text{note } T_2 = 0.$$

We first use Equation (10.19) to determine the constants A_n.

$$A_n = \frac{2}{L}\int_0^L T_1\left(\frac{L - x}{L}\right)^2 \sin\frac{\pi n x}{L}\, dx - \frac{2}{\pi n}T_1$$

or, after the required calculus, we find

$$A_n = \frac{4T_1}{(\pi n)^3}(\cos \pi n - 1).$$

Now, substitution of this expression into Equation (10.18) yields our desired results.

$$u(x; t) = \sum_{n=1}^{\infty}\left[\frac{4T_1}{(\pi n)^3}(\cos \pi n - 1)e^{-\alpha(\pi n/L)^2 t} + \frac{2T_1}{\pi n}\right]\sin\frac{\pi n x}{L}.$$

In Figure 10.6 we plot this equation (with 450 terms) for various equally spaced increments of time. Note the ringing, or Gibb's effect, at the discontinuity $x = 0$.

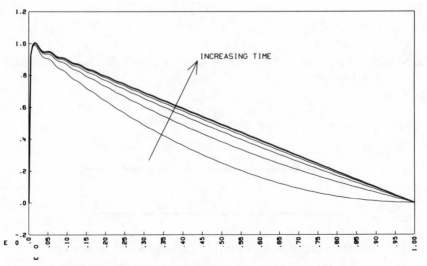

FIGURE 10.6. Example of temperature profile at various times.

Ends Held at Specified Variable Temperatures

Next we examine the more general situation in which the ends of the finite body are allowed to vary with some prescribed functions of time. That is, we permit the temperature at one end ($x = 0$) to vary according to the function $\varphi_1(t)$ and the other end according to the function $\varphi_2(t)$. The initial temperature profile of the body is again described by the function $f(x)$. This situation is illustrated in Figure 10.7. Mathematically, the initial and boundary conditions for this problem are given as

$$u(x;0) = f(x), \qquad [f(0) = \varphi_1(t), f(L) = \varphi_2(t)]$$

$$u(0;t) = \varphi_1(t),$$

$$u(L;t) = \varphi_2(t).$$

As usual, we attack this problem by first taking the finite Fourier sine transform of both sides of Equation (10.11) which results in

$$\frac{dU_S(n;t)}{dt} = -\alpha \frac{\pi n}{L} \left[\frac{\pi n}{L} U_S(n;t) + \frac{2}{L}(-1)^n u(L;t) - \frac{2}{L} u(0;t) \right],$$

$$n = 1,2,\ldots.$$

Next we apply the boundary conditions and a little algebra to obtain

$$(10.20) \quad \frac{dU_S(n;t)}{dt} + \alpha \left(\frac{\pi n}{L} \right)^2 U_S(n;t) = \frac{2\alpha\pi n}{L^2} \left[\varphi_1(t) - (-1)^n \varphi_2(t) \right]$$

$$n = 1,2,\ldots.$$

Equations (10.20) constitute a set of general first-order linear differential equations of the form

$$\frac{dy(t)}{dt} + ky(t) = g(t).$$

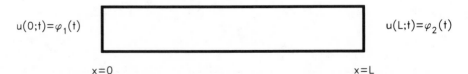

$u(0;t)=\varphi_1(t)$ $u(L;t)=\varphi_2(t)$

x=0   x=L

FIGURE 10.7.   Finite body with ends at specified variable temperatures.

The classical complete solution to this type of equation is given as

$$y(t) = Ae^{-kt} + e^{-kt}\int_0^t g(\tau)e^{k\tau}\,d\tau.$$

If we now let

$$k = \alpha\left(\frac{\pi n}{L}\right)^2$$

and

$$g(t) = \frac{2\alpha\pi n}{L^2}\left[\varphi_1(t) - (-1)^n\varphi_2(t)\right] \qquad n = 1, 2, \ldots,$$

then the general solution to Equation (10.20) can be directly written as

$$U_S(n; t) = A_n e^{-\alpha(\pi n/L)^2 t}$$

$$+ \frac{2\alpha\pi n}{L^2}e^{-\alpha(\pi n/L)^2 t}\int_0^t\left[\varphi_1(\tau) - (-1)^n\varphi_2(\tau)\right]e^{\alpha(\pi n/L)^2 \tau}\,d\tau$$

or, back in the spatial domain, we have

$$(10.21) \quad u(x; t) = \sum_{n=1}^{\infty} A_n e^{-\alpha(\pi n/L)^2 t}\sin\frac{\pi nx}{L} + \sum_{n=1}^{\infty}\frac{2\alpha\pi n}{L^2}e^{-\alpha(\pi n/L)^2 t}$$

$$\times \int_0^t\left[\varphi_1(\tau) - (-1)^n\varphi_2(\tau)\right]e^{\alpha(\pi n/L)^2 \tau}\,d\tau\sin\frac{\pi nx}{L}.$$

Invoking the initial condition $u(x; 0) = f(x)$ we find

$$u(x; 0) = f(x) = \sum_{n=1}^{\infty} A_n\sin\frac{\pi nx}{L}$$

which (as already illustrated) implies

$$(10.22) \qquad A_n = \frac{2}{L}\int_0^L f(x)\sin\frac{\pi nx}{L}\,dx.$$

As in the previous example problem, we note that if $\varphi_1(t) \neq 0$, then there will be a discontinuity at $x = 0$ and, consequently, $u(0; t)$ must be determined as a right-hand limit. Also because of this discontinuity the solution of Equation (10.21) will exhibit artificial ringing, or Gibb's effect. Similar remarks hold at $x = L$ if $\varphi_2(t) \neq 0$. As a specific example, let us suppose that

$$f(x) = 0,$$

$$\varphi_2(t) = 0,$$

and

$$\varphi_1(t) = \begin{cases} T_1, & t \geqslant 0 \\ 0, & t < 0. \end{cases}$$

Because $f(x) = 0$, we know from Equation (10.22) that $A_n = 0$ for all n. Thus Equation (10.21) becomes

$$u(x; t) = \sum_{n=1}^{\infty} \frac{2\alpha\pi n}{L^2} e^{-\alpha(\pi n/L)^2 t} \int_0^t T_1 e^{\alpha(\pi n/L)^2 \tau} \, d\tau \sin \frac{\pi n x}{L}$$

or, performing the integration within this summation, we find

$$(10.23) \qquad u(x; t) = \sum_{n=1}^{\infty} \frac{2T_1}{\pi n} \left[1 - e^{-\alpha(\pi n/L)^2 t} \right] \sin \frac{\pi n x}{L}.$$

Shown in Figure 10.8 is a plot of Equation (10.23) (with $n = 450$) for various equally spaced increments in time.

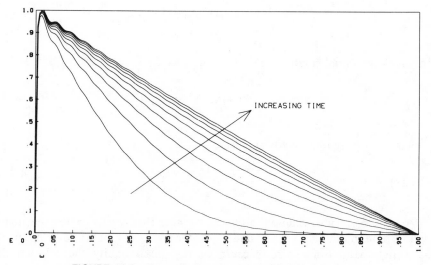

FIGURE 10.8. Example of temperature profile at various times.

Finite Body With Both Ends Insulated

In this section we examine the situation in which both ends of the finite body are insulated to prevent the flow of heat across them. That is, $\partial u(x; t)/\partial x = 0$ at $x = 0$ and $x = L$. Let us also assume that the initial temperature distribution of the body is described as $f(x)$. These boundary and initial conditions can be mathematically stated as

$$u'(0; t) = 0,$$

$$u'(L; t) = 0,$$

$$u(x; 0) = f(x), \qquad f'(0) = f'(L) = 0.$$

Inasmuch as the boundary conditions of this problem are specified in terms of the derivative of $u(x; t)$, we solve this problem using the finite Fourier cosine transform. To do so we must multiply both sides of Equation (10.11) by

$$\frac{2}{L} \cos \frac{\pi n x}{L}$$

and integrate the resulting expressions from 0 to L:

$$\frac{2}{L} \int_0^L \frac{\partial u(x; t)}{\partial t} \cos \frac{\pi n x}{L} \, dx = \alpha \frac{2}{L} \int_0^L \frac{\partial^2 u(x; t)}{\partial x^2} \cos \frac{\pi n x}{L} \, dx.$$

Using reasoning analogous to that which we applied earlier in the chapter to the sine transform we obtain

(10.24)

$$\frac{dU_C(n; t)}{dt} = \alpha \left[-\left(\frac{\pi n}{L} \right)^2 U_C(n; t) + \frac{2}{L} (-1)^n u'(L; t) - \frac{2}{L} u'(0; t) \right],$$

$$n = 1, 2, \ldots .$$

NOTE: From Equation (10.9) we can see that when $n = 0$ we must divide the right-hand side of the preceding equation by two to obtain the proper results. Using the prevailing boundary conditions in Equation (10.24) we obtain

$$\frac{dU_C(n; t)}{dt} + \alpha \left(\frac{\pi n}{L} \right)^2 U_C(n; t) = 0, \qquad n = 0, 1, \ldots .$$

As we have already indicated, the solution to this differential equation is

$$U_C(n; t) = A_n e^{-\alpha(\pi n/L)^2 t}, \qquad n = 0, 1, \ldots .$$

Now, using Equation (10.8) we go from the transform domain back to the spatial domain:

$$(10.25) \qquad u(x; t) = A_0 + \sum_{n=1}^{\infty} A_n e^{-\alpha(\pi n/L)^2 t} \cos \frac{\pi n x}{L}.$$

Finally, applying the initial conditions ($t = 0$) we find

$$u(x; 0) = f(x) = A_0 + \sum_{n=1}^{\infty} A_n \cos \frac{\pi n x}{L},$$

from which we see that the constants A_n are the Fourier series coefficients of the (assumed) even function $f(x)$:

$$A_0 = \frac{1}{L} \int_0^L f(x)\, dx,$$

$$(10.26)$$

$$A_n = \frac{2}{L} \int_0^L f(x) \cos \frac{\pi n x}{L}\, dx.$$

TWO-DIMENSIONAL STEADY STATE HEAT FLOW

In this section we examine a finite two-dimensional body in which the heat transfer has reached steady state conditions. That is, the temperature variation within the body is no longer a function of time but instead only a function of position. Mathematically, we have

$$\frac{\partial u(x; t)}{\partial t} = 0.$$

When this condition is used in Equation (10.3) we obtain

$$(10.27) \qquad \frac{\partial^2 u(x, y)}{\partial x^2} + \frac{\partial^2 u(x, y)}{\partial y^2} = 0$$

which is known as the Laplace equation.

Consider the rectangular plate shown in Figure 10.9 in which the temperature along the top is maintained at T_1 whereas the other three sides have their temperatures fixed at zero.

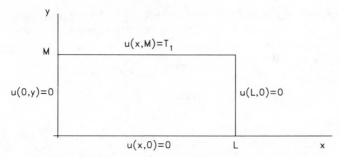

FIGURE 10.9. Rectangular plate with sides at fixed temperature.

Mathematically summarizing the boundary conditions we have

$$u(0, y) = 0,$$

$$u(L, y) = 0,$$

$$u(x, 0) = 0,$$

$$u(x, M) = T_1.$$

We obtain a solution to this problem in two different ways. First we use a successive transform technique in which we take the finite sine transform of Equation (10.27) with respect to x and then take the finite sine transform of the resulting expression with respect to y. Transforming Equation (10.27) with respect to x we obtain

$$\frac{-\pi n}{L}\left[\frac{\pi n}{L}U_S(n, y) + \frac{2}{L}(-1)^n u(L, y) - \frac{2}{L}u(0, y)\right] + \frac{\partial^2 U_S(n, y)}{\partial y^2} = 0.$$

When the proper boundary conditions are substituted into this equation we obtain

(10.28) $$\frac{\partial^2 U_S(n, y)}{\partial y^2} - \left(\frac{\pi n}{L}\right)^2 U_S(n, y) = 0.$$

We now have a second-order differential equation in terms of the "new function" $U_S(n, y)$. To solve this one we take the finite sine transform of Equation (10.28) with respect to the variable y.

$$\frac{\pi m}{M}\left[\frac{\pi m}{M}\hat{U}_S(n, m) + \frac{2}{M}(-1)^m U_S(n, M) - \frac{2}{M}U_S(n, 0)\right]$$

$$+ \left(\frac{\pi n}{L}\right)^2 \hat{U}_S(n, m) = 0,$$

where $\hat{U}_S(n, m)$ is the finite sine transform of $U_S(n, y)$ with respect to the variable y. Note that in this equation we require the boundary conditions in terms of the transform $U_S(n, y)$ rather than $u(x, y)$. Therefore, we proceed to calculate them as follows:

$$(10.29) \quad U_S(n,0) = \frac{2}{L} \int_0^L u(x,0) \sin \frac{\pi n x}{L} \, dx = 0,$$

$$U_S(n, M) = \frac{2}{L} \int_0^L u(x, M) \sin \frac{\pi n x}{L} \, dx = \frac{2}{L} \int_0^L T_1 \sin \frac{\pi n x}{L} \, dx,$$

or

$$(10.30) \quad U_S(n, M) = \frac{2T_1}{\pi n} (1 - \cos \pi n).$$

Substituting these expressions into the previous equation we obtain

$$\left[\left(\frac{\pi n}{L} \right)^2 + \left(\frac{\pi m}{M} \right)^2 \right] \hat{U}_S(n, m) = \frac{4T_1}{\pi n M} (-1)^m (\cos \pi n - 1),$$

or

$$\hat{U}_S(n, m) = \frac{4T_1(-1)^m}{\pi n M} (\cos \pi n - 1) \left[\frac{1}{(\pi n/L)^2 + (\pi m/M)^2} \right].$$

Thus inverse transforming this equation with respect to m we find

$$U_S(n, y) = \sum_{m=1}^{\infty} \frac{4T_1(-1)^m (\cos \pi n - 1)}{\pi n M \left[(\pi n/L)^2 + (\pi m/M)^2 \right]} \sin \frac{\pi m y}{M}.$$

Now inverse transforming this equation with respect to n we arrive at

(10.31)

$$u(x, y) = \sum_{n=1}^{\infty} \sum_{m=1}^{\infty} \frac{4T_1(-1)^m (\cos \pi n - 1)}{\pi n M \left[(\pi n/L)^2 + (\pi m/M)^2 \right]} \sin \frac{\pi m y}{M} \sin \frac{\pi n x}{L}.$$

Equation (10.31) describes the spatial temperature distribution within the body. We now obtain another equivalent solution to this problem. We begin with the second-order differential equation (10.28) whose solution is well-known and given as

$$(10.32) \qquad\qquad U_S(n, y) = Ae^{-\pi n y/L} + Be^{\pi n y/L}.$$

We solve for the constants A and B by applying the boundary conditions:

$$0 = A + B,$$

$$\frac{2T_1}{\pi n}(1 - \cos \pi n) = Ae^{-\pi n M/L} + Be^{\pi n M/L},$$

which yields

$$A = -B = \frac{2T_1(1 - \cos \pi n)/\pi n}{e^{\pi n M/L} - e^{-\pi n M/L}}.$$

When these values are substituted into Equation (10.32) we obtain

$$U_S(n, y) = \left(\frac{e^{\pi n y/L} - e^{-\pi n y/L}}{e^{\pi n M/L} - e^{-\pi n M/L}} \right) \frac{2T_1}{\pi n}(1 - \cos \pi n),$$

but since

$$\sinh z = \frac{e^z - e^{-z}}{2},$$

we can rewrite this equation in a more simplified form:

$$U_S(n, y) = \left[\frac{\sinh(\pi n y/L)}{\sinh(\pi n M/L)} \right] \frac{2T_1}{\pi n}(1 - \cos \pi n).$$

Now, inverse transforming the preceding equation (with respect to x) back into the spatial domain we find

$$(10.33) \quad u(x, y) = \sum_{n=1}^{\infty} \left[\frac{\sinh(\pi n y/L)}{\sinh(\pi n M/L)} \right] \frac{2T_1}{\pi n}(1 - \cos \pi n) \sin \frac{\pi n x}{L}.$$

Both Equations (10.33) and (10.31) describe the spatial temperature distribution in the rectangular plate shown in Figure 10.9. Equation (10.31) is a series solution in both the x and y dimension, whereas Equation (10.33) provides a series solution in the x dimension and a closed form solution in the y dimension.

TWO-DIMENSIONAL TRANSIENT HEAT FLOW

In the previous section we examined a simple two-dimensional heat transfer problem in which steady state conditions had been achieved. In this section we solve a transient, or nonsteady state, two-dimensional problem. To be more specific, we consider the finite body shown in Figure 10.10 which has all four edges maintained at zero temperature and an initial thermal profile described by the function $f(x, y)$. Mathematically, these conditions may be expressed as:

$$u(x, 0; t) = 0,$$

$$u(x, M; t) = 0,$$

$$u(0, y; t) = 0,$$

$$u(L, y; t) = 0,$$

$$u(x, y; 0) = f(x, y), \quad f(0, y) = f(L, y) = f(x, 0) = f(x, M) = 0.$$

The dynamic behavior of this system is governed by differential Equation (10.3). Our approach to solving this problem is to first take the finite sine transform of both sides of this equation with respect to the variable x and apply the boundary conditions

$$u(0, y; t) = u(L, y; t) = 0.$$

This move results in the following equation:

(10.34)

$$\frac{\partial^2 U_s(n, y; t)}{\partial y^2} - \alpha \left(\frac{\pi n}{L} \right)^2 U_s(n, y; t) = \frac{\partial U_s(n, y; t)}{\partial t}, \quad n = 1, 2, \ldots .$$

where $U_s(n, y; t) = S[u(x, y; t)]$.

Equation (10.34) is a partial differential equation in terms of the quantity $U_s(n, y; t)$. Therefore, we again take the finite sine transform of both sides

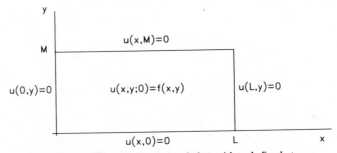

FIGURE 10.10.   Two-dimensional plate with ends fixed at zero.

with respect to the variable y and apply the boundary conditions

$$U_s(n,0; t) = \frac{2}{L} \int_0^L u(x,0; t)\sin\frac{\pi nx}{L} dx = 0,$$

$$U_s(n, M; t) = \frac{2}{L} \int_0^L u(x, M; t)\sin\frac{\pi nx}{L} dx = 0, \qquad n = 1, 2, \dots .$$

The resulting equation is

(10.35)

$$-\alpha\left[\left(\frac{\pi n}{L}\right)^2 + \left(\frac{\pi m}{M}\right)^2\right]\hat{U}_s(n, m; t) = \frac{\partial\hat{U}_s(n, m; t)}{\partial t}, \qquad \begin{matrix} n = 1, 2, \dots, \\ m = 1, 2, \dots, \end{matrix}$$

where $\hat{U}_s(n, m; t) = S[U_s(n, y; t)]$. Equation (10.35) is a first-order linear differential equation and, as we have already demonstrated several times, its solution is given by

(10.36) $\hat{U}_s(n, m; t) = A_{mn}e^{-\alpha[(\pi n/L)^2 + (\pi m/M)^2]t}, \qquad \begin{matrix} n = 1, 2, \dots, \\ m = 1, 2, \dots, \end{matrix}$

At this point we have our solution in terms of the double transform $\hat{U}_s(n, m; t)$. To obtain the answer in terms of $u(x, y; t)$ we must perform a double inverse transform. First with respect to y:

$$U_s(n, y; t) = \sum_{m=1}^{\infty} A_{mn}e^{-\alpha[(\pi n/L)^2 + (\pi m/M)^2]t}\sin\frac{\pi my}{M}, \qquad n = 1, 2, \dots .$$

Then with respect to x:

(10.37) $u(x, y; t) = \sum_{n=1}^{\infty} \sum_{m=1}^{\infty} A_{mn}e^{-\alpha[(\pi n/L)^2 + (\pi m/M)^2]t}\sin\frac{\pi my}{M} \sin\frac{\pi nx}{L}.$

Applying the initial conditions ($t = 0$) in the preceding equation, we obtain

(10.38) $u(x, y; 0) = f(x, y) = \sum_{N=1}^{\infty} \sum_{m=1}^{\infty} A_{mn}\sin\frac{\pi my}{M} \sin\frac{\pi nx}{L}.$

This equation is a two-dimensional Fourier series representation of the (assumed odd) function $f(x, y)$. Therefore,

(10.39) $A_{mn} = \frac{16}{LM} \int_0^M \int_0^L f(x, y)\sin\frac{\pi nx}{L} \sin\frac{\pi my}{M} dx \, dy.$

HEAT FLOW IN AN INFINITE ONE-DIMENSIONAL BODY

In this section we examine the heat flow or temperature distribution in a body that is, for all practical purposes, infinite in its dimensions. By this we mean that the body in question is sufficiently large so that the time required for heat to flow from one end to the other is many times greater than the time required for the heat flow to stabilize over the region of interest. Another way to look at this is that the behavior at the boundaries of the body does not affect the heat conduction within. In other words, there are no boundary conditions imposed upon this type of problem. For simplicity, we restrict our attention to one dimension, in which case the governing differential equation becomes

$$(10.40) \qquad \frac{\partial u(x; t)}{\partial t} = \alpha \frac{\partial^2 u(x; t)}{\partial x^2}.$$

We will solve this equation using an approach that is similar to the one we used in Chapter 7 to analyze the vibrational motion of an infinite string. That is, assume that the solution is a separable function in both space and time:

$$(10.41) \qquad u(x; t) = \psi(x)\varphi(t).$$

Furthermore, let us assume a particular form for the spatial portion, namely,

$$(10.42) \qquad \psi(x) = e^{2\pi i w x}.$$

As in Chapter 7, we use methods of elementary calculus to obtain

$$\frac{d\psi(x)}{dx} = 2\pi i w e^{2\pi i w x} \quad \text{and} \quad \frac{d^2\psi(x)}{dx^2} = -4\pi^2 w^2 e^{2\pi i w x}.$$

Therefore [Equation (10.41)], we have

$$\frac{\partial u(x; t)}{\partial t} = \frac{d\varphi(t)}{dt} e^{2\pi i w x} \quad \text{and} \quad \frac{\partial^2 u(x; t)}{\partial x^2} = -4\pi^2 w^2 e^{2\pi i w x}\varphi(t).$$

When these two expressions are substituted back into Equation (10.40) and the $\exp(2\pi i w x)$ term cancelled, we obtain

$$\frac{d\varphi(t)}{dt} + 4\pi^2 w^2 \alpha \varphi(t) = 0.$$

The preceding first-order differential equation is easily solved to yield

$$(10.43) \qquad \varphi(t) = A e^{-4\pi^2 w^2 \alpha t},$$

where A is a constant to be determined by the initial conditions of the particular problem. Combining Equations (10.41), (10.42), and (10.43) we find

$$(10.44) \qquad u(x; t) = Ae^{-4\pi^2 \alpha t w^2} e^{2\pi i w x}.$$

If we now consider this infinite body to be a system with spatial input $e(x) = \exp(2\pi i w x)$ and resulting output $u(x; t)$, then Equation (10.44) tells us that $e(x)$ is an eigenfunction of this system and its transfer function is given by

$$(10.45) \qquad H_t(w) = Ae^{-4\pi^2 \alpha t w^2}$$

which is valid for any and all times t.

Using this transfer function we are able to determine the output of this system when the input is an arbitrary spatial function $f(x)$ [with Fourier transform $F(w)$]. That is, in the frequency domain we write

$$(10.46) \qquad U(w; t) = H_t(w)F(w) = Ae^{-4\pi^2 \alpha t w^2} F(w),$$

where $U(w; t)$ is the Fourier transform of $u(x; t)$ with respect to x. Based on the remarks in Chapter 6, we know that the (spatial) impulse response of this system is simply the inverse Fourier transform (with respect to w) of the transfer function $H_t(w)$ and is given as [see Equation (3.16)]

$$(10.47) \qquad h_t(x) = A \frac{1}{\sqrt{4\alpha\pi t}} e^{-x^2/4\pi t}.$$

In the spatial domain (convolution integral) Equation (10.46) becomes

$$(10.48) \qquad u(x; t) = A \frac{1}{\sqrt{4\alpha\pi t}} \int f(\xi) e^{-(x-\xi)^2/4\pi t} \, d\xi.$$

Inasmuch as this equation is valid for any time t, let us assume that at $t = 0$ the temperature distribution in the body is described as $u(x; 0) = f(x)$. Thus Equation (10.48) may be considered to describe the temperature profile in the body as a function of increasing time. As a specific example, let us assume that at $t = 0$ we impress upon this body a (spatial) delta distribution centered at $\xi = 0$. In this case Equation (10.48) simply reverts back to the impulse response as per Equation (10.47). This is shown in Figure 10.11 for various values of time. Note that in the limit as t approaches zero this impulse response function becomes a spatial delta function.

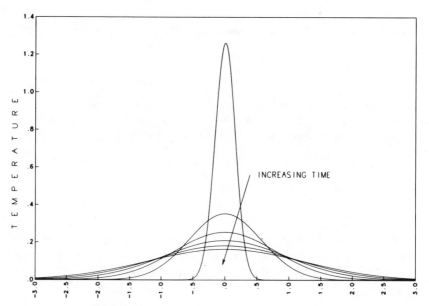

FIGURE 10.11. Impulse response of infinite medium.

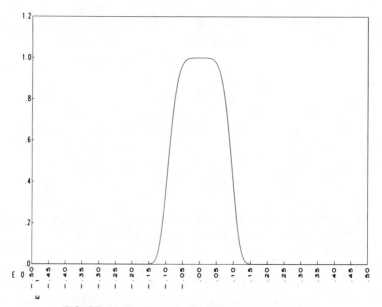

FIGURE 10.12. Example of initial temperature profile.

318

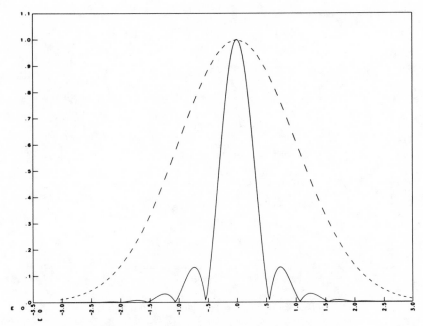

FIGURE 10.13. Transform of input and transfer function.

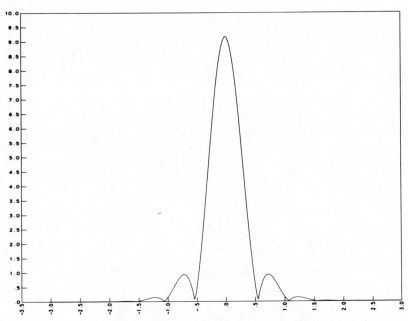

FIGURE 10.14. Fourier transform of output.

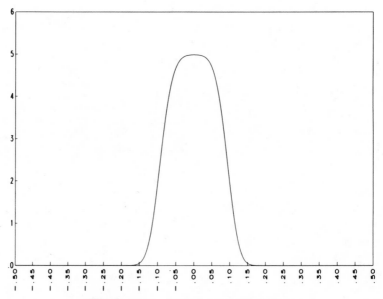

FIGURE 10.15. Temperature profile at time *t*.

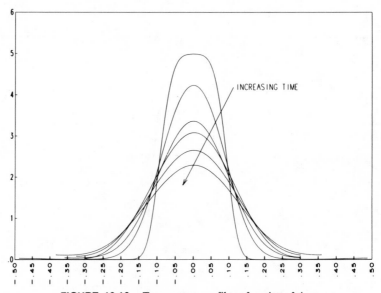

INCREASING TIME

FIGURE 10.16. Temperature profile as function of time.

Next let us assume that the initial profile is described by the function

$$u(x;0) = f(x) = e^{-|x|^5},$$

shown in Figure 10.12. We solve this problem (numerically on the computer) in the frequency domain using Equation (10.46). Shown in Figure 10.13 is (absolute value of) the Fourier transform of $f(x)$ along with the transfer function of Equation (10.45) (dashed line) for a specific (early) value of t. In Figure 10.14 we show the resulting product, and in Figure 10.15 the inverse Fourier transform of this product (i.e., the temperature profile in the body at this time t). In Figure 10.16 we show the results of this same analysis performed for several (increasing) values of time.

SUMMARY

In this chapter we have shown how Fourier analysis can be used to facilitate the solution of the heat, or diffusion, equation. We began the chapter with a brief derivation of this equation and then introduced the finite Fourier sine and cosine transforms. As it turned out, these transforms were particularly useful in the study of heat flow in finite bodies. The remainder of the chapter was devoted to presenting illustrative solutions to several classes of heat transfer problems. Although it was obviously impossible to consider all interesting cases, an understanding of those presented in this chapter should enable the reader to set up and solve many others.

BIBLIOGRAPHY

Raven, F. H., *Mathematics of Engineering Systems*. McGraw-Hill, New York, 1966.

Churchill, R. V., *Fourier Series and Boundary Value Problems*. McGraw-Hill, New York, 1941.

Tolstov, G. P., *Fourier Series*. Prentice-Hall, Englewood Cliffs, N.J., 1962.

CHAPTER

11

STOCHASTIC ANALYSIS

In Chapter 6 we discussed the concept of a system and its transfer function. We saw that when the input to a system was a deterministic signal (i.e., could be described analytically), we were able to use this transfer function to obtain an analytical expression for the output. In this chapter we examine the situation in which the input signal (and, consequently, the output) is stochastic, or random, in nature. These random signals are characterized in terms of their probabilistic or statistical properties. Therefore, we begin this chapter with a brief discussion of statistics, probability, and correlation theory.

REVIEW OF STATISTICS

In this section we present those concepts from statistics that are useful to us in the remainder of the chapter.

In very general terms, *Statistics may be defined as the science of collecting, classifying, presenting, and interpreting numerical data.* Associated with this science are the following terms.

POPULATION. A collection, or set, of individuals, objects, or measurements whose properties are to be analyzed.

323

SAMPLE.   A subset of the population.

VARIABLE.   A characteristic or phenomenon about each individual element of the population or sample.

DATA.   The value of the variable (i.e., numbers or measurements collected as a result of an observation).

EXPERIMENT.   A planned activity whose results yield a set of data.

PARAMETER.   A measurable characteristic of the entire population.

STATISTIC.   A measurable characteristic of a sample.

In most of our work we assume that the data were obtained from a population.

For example, suppose that in some well-known university there are 50 students in a Fourier analysis class. Furthermore, let us assume that on one particular day the instructor asked each student in the class to count the money on his or her person and record that amount to the nearest dollar. The results of that experiment are shown in Table 11.1. In this example we consider the population to be the set of all individuals in the Fourier analysis class. The experiment was performed by the instructor when the class was asked to count their money and the variable in question is the amount of money each student possessed. The resulting data is that displayed in Table 11.1. If we were to calculate the average value of the data in this table, then we would obtain a parameter. Now let us assume that only 10 students in the class are chosen (randomly) and asked to repeat the experiment. In this case these 10 students would constitute a sample and the average number of dollars calculated would be called a statistic.

Data such as that presented in Table 11.1 is called *raw data* because it appears in the form in which it was collected. Raw data is difficult, if not impossible, to interpret and, therefore, we must present it in some other more useful or systematic form. One such form is known as a frequency distribution which is obtained by listing all possible data values x_j in ascending (or

TABLE 11.1.   Example of Raw Data

$1	21	36	19	20	5	14	6	29	23
22	10	27	24	10	21	28	18	8	13
39	26	17	28	45	32	16	31	12	16
29	20	25	24	22	47	35	23	25	58
14	41	54	3	43	37	30	10	49	32

TABLE 11.2. Example of a Frequency Distribution

x_j	f_j	x_j	f_j	x_j	f_j
0	0	20	2	40	0
1	1	21	2	41	1
2	0	22	2	42	0
3	1	23	2	43	1
4	0	24	2	44	0
5	1	25	2	45	1
6	1	26	1	46	0
7	0	27	1	47	1
8	1	28	2	48	0
9	0	29	2	49	1
10	3	30	1	50	0
11	0	31	1	51	0
12	1	32	2	52	0
13	1	33	0	53	0
14	2	34	0	54	1
15	0	35	1	55	0
16	2	36	1	56	0
17	1	37	1	57	0
18	1	38	0	58	1
19	1	39	1	59	0

descending) order and then counting the number of times or frequency f_j that each value occurs. Shown in Table 11.2 is the frequency distribution of the raw data presented in Table 11.1. We often use a *relative frequency distribution* which is simply a frequency distribution in which each f_j has been divided, or normalized, by the total number of data values N. The most convenient way to "read" this distribution is to construct some form of graph which plots frequency f_j versus data value x_j. This is illustrated in Figure 11.1.

The *total cumulative frequency distribution* of a set of data is defined as

$$\hat{F}_j = \sum_{k=1}^{j} f_k, \quad j = 1, \ldots, M.$$

From this equation we can easily see that $\hat{F}_j$ is the number of data values between x_1 and x_j, inclusively. Also this equation reveals that $\hat{F}_j - \hat{F}_k$ is the number of data values from x_k to x_j ($j > k$). In our work we are mainly concerned with the (*relative*) *cumulative frequency distribution*[†] given as

$$F_j = \sum_{k=1}^{j} \frac{f_k}{N} = \frac{\hat{F}_j}{N}.$$

[†]From here on when we use the term cumulative frequency distribution we mean the relative cumulative frequency distribution unless specifically stated otherwise.

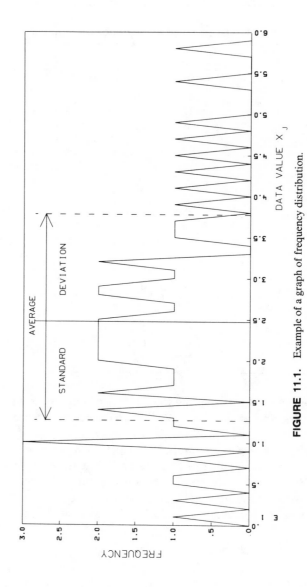

FIGURE 11.1. Example of a graph of frequency distribution.

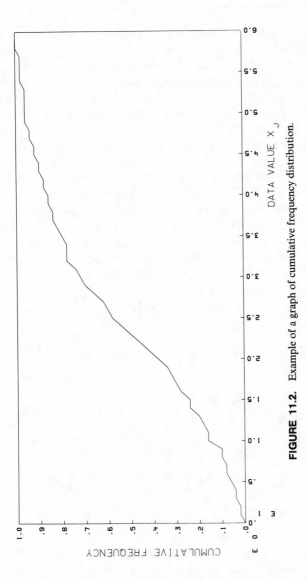

FIGURE 11.2. Example of a graph of cumulative frequency distribution.

327

This distribution is simply the ratio of the number of data values between x_1 and x_j to the total number of data points. Also $F_j - F_k$ is the fraction of the total number of data values that lie between x_k and x_j ($j > k$). Shown in Figure 11.2 is a plot of the cumulative frequency distribution of the data presented in Table 11.1.

Another interpretation of the cumulative frequency distribution is that of a ranking system for the data values x_j. That is, associated with each value x_j is a number F_j which is the fraction of the total number of data points below x_j.

Although a graphical display of a frequency, or cumulative frequency, distribution is often useful to obtain an intuitive or physical "feeling" for the scatter of the data, it does not permit a quantitative description so that one set of data can be compared to another. Perhaps the most obvious quantitative description of a set of data or frequency distribution is one that measures the central tendency of the distribution. There are several such measures; however, the most useful is the mean or average (denoted as μ). This is simply the sum of the values x_j divided by their total number N:

(11.1)
$$\mu = \sum_{i=1}^{N} \frac{x_i}{N}$$

or, in terms of the frequency of occurrence of the values,

(11.2)
$$\mu = \frac{\sum_{j=1}^{M} f_j x_j}{\sum_{j=1}^{M} f_j}$$

where, obviously,

$$\sum_{j=1}^{M} f_j = N.$$

In addition to the mean of a frequency distribution we would also like to have some quantitative measure of the dispersion of the data values about this mean. For example, the two sets of data shown schematically in Figure 11.3 both have the same mean value. However, in one case the data values are closely grouped around the mean, whereas in the other they are widely dispersed. Just as with the measure of central tendency, there are several measures of dispersion. Intuitively, the most pleasing is the mean deviation, defined as

$$\text{MD} = \sum_{i=1}^{N} \frac{|\mu - x_i|}{N}.$$

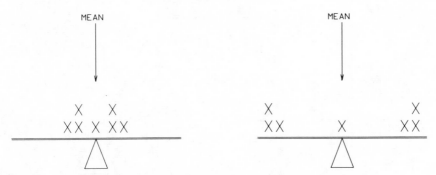

FIGURE 11.3. Example of a same mean but different dispersions.

We note that the expression $|\mu - x_i|$ is the distance from the mean to the data value x_i. Thus the equation tells us that the mean deviation is the average distance of all the data values from the mean μ. The trouble with this measure of dispersion is that it defines distance in terms of the absolute value norm which does not easily lend itself to algebraic manipulations. If, instead, we define the distance of the value x_i from the mean μ in terms of the square norm $(\mu - x_i)^2$ then we obtain a measure of dispersion known as the variance[†] (denoted as σ^2):

$$(11.3) \qquad \sigma^2 = \sum_{i=1}^{N} \frac{(\mu - x_i)^2}{N}$$

or, in terms of the frequency of occurrence f_j,

$$(11.4) \qquad \sigma^2 = \frac{\sum_{j=1}^{M} (\mu - x_j)^2 f_j}{\sum_{j=1}^{M} f_j}$$

The *standard deviation* (SD) of a frequency distribution is defined as the positive square root of the variance (i.e., SD = σ). This is the most commonly used measure of dispersion.

We now have a quantitative measure of both the central tendency and dispersion of a frequency distribution. Our next task is to obtain a measure of how an individual data value (or score) ranks in such a distribution. For example, a score of 85 on a calculus exam is difficult to interpret (good, fair, or poor) without knowing how well the rest of the class performed on the same

---

[†] This definition of the variance assumes that we are dealing with a population. When a sample is used the denominator becomes $N - 1$ rather than N.

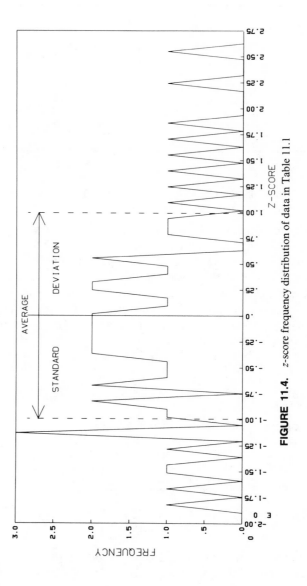

FIGURE 11.4. z-score frequency distribution of data in Table 11.1

exam. However, if we know that the class mean was 75, then we can relate this 85 score to the average performance of the class; that is, it is 10 points above the mean. Now if we are also given that the standard deviation of the scores was 5, then we can obtain an even better idea of the rank of our 85 score. That is, it is two standard deviations above the mean.

This procedure of expressing data values in terms of the number of standard deviations above or below the mean is known as conversion to z-scores. More formally, a data value x_j is converted to a z-score z_j as per the equation

(11.5)
$$z_j = \frac{x_j - \mu}{\sigma}.$$

When all the scores in a frequency distribution are converted to z-scores we obtain a new frequency distribution. This new distribution has the following unique properties: (1) its mean is zero and (2) its standard deviation is unity. Shown in Figure 11.4 is the z-score frequency distribution of the data presented in Table 11.1. This should be compared to the frequency distribution plot of Figure 11.1. Although the plots appear identical, note the values of the means and standard deviations.

Another useful feature of z-scores is that they can be used to compare scores or data values from different distributions. For example, suppose we wish to compare a score of 63 on a math exam with a score of 80 on an english exam. Also let us assume that for the math exam the mean $\mu = 60$ and the standard deviation $\sigma = 1.3$, whereas for the english exam, $\mu = 83$ and $\sigma = 3$. The z-scores for the respective exams are: math $z = 2.31$ and english $z = -1.0$. Therefore, if both instructors "grade on a curve," the 63 on the math exam can be considered the better performance.

REVIEW OF PROBABILITY THEORY

We now turn our attention to the concept of probability and present those topics of use to us in this chapter. The *probability that an event will occur is defined as the relative frequency with which that event occurs or the relative frequency with which that event can be expected to occur.* The probability or relative frequency of an event's occurrence can often be obtained either experimentally or analytically. We first discuss analytical means. Let us denote the population space of all possible outcomes of our probability experiment as S. Also let us assume that every outcome in S has an equally likely chance of occurring. Then the probability of an event A occurring is the ratio of the number of outcomes $n(A)$ in S that satisfy the definition of event A to the total number of points $n(S)$ in the space:

(11.6)
$$p(A) = \frac{n(A)}{n(S)}.$$

Example

In a single toss of a coin, what is the probability of obtaining a head?

Solution. The first task in the solution of *any* probability problem is to define the population space S. In this particular case it is a rather simple matter, either heads or tails; thus

$$S = \{H, T\} \qquad n(S) = 2.$$

In this space it is rather obvious that event A has only one way of occurring and, therefore, $n(A) = 1$. Thus $p(A) = \frac{1}{2}$.

Example

Two coins are tossed simultaneously. What is the probability that at least one head shows?

Solution. In this case the population space will be a set of order pairs (X, Y) in which X is the result on the first coin and Y is the result on the second:

$$S = \{(H, H), (H, T), (T, H), (T, T)\}, \qquad n(S) = 4.$$

In this space we see that event A is satisfied by three of the outcomes; that is, $n(A) = 3$. Thus $p(A) = \frac{3}{4}$.

Example

Let two dice be rolled together. What is the probability that the sum showing will be equal to seven?

Solution. To determine the population space S of all possible outcomes for this problem we construct the matrix type sketch shown in Figure 11.5. Along the top is the result on the first die and along the side the result on the other. The elements in the matrix grid show the sums of the 36 possible outcomes $[n(S) = 36]$. We see that six satisfy the definition of event A. Thus $n(A) = 6$ and $p(A) = \frac{6}{36} = \frac{1}{6}$.

To experimentally determine the probability of an event A occurring we simply perform an experiment n times and count the number of times that event A actually occurred $[\#(A)]$. The (experimental) probability of the event A occurring, denoted as $P'(A)$, is defined as

$$(11.7) \qquad\qquad P'(A) = \frac{\#(A)}{n}.$$

The law of large numbers states: *As the number of times that an experiment is repeated is increased, the ratio of the number of successful occurrences to the*

DIE 1 $\longrightarrow$

	1	2	3	4	5	6
1	2	3	4	5	6	7
2	3	4	5	6	7	8
3	4	5	6	7	8	9
4	5	6	7	8	9	10
5	6	7	8	9	10	11
6	7	8	9	10	11	12

DIE 2

FIGURE 11.5. Example of a probability problem.

number of trials will tend to approach the analytical probability of the outcome for an individual trial. Thus we can write

$$\lim_{n \to \infty} P'(A) = p(A).$$

Based upon the way in which $p(A)$ is defined, it should be rather obvious that $0 \leqslant p(A) \leqslant 1$. A probability of zero means that event A cannot happen (e.g., the sum on the roll of two dice equals one), whereas a probability of one means that the event is certain to happen (e.g., the sum of the roll of two dice is less than or equal to 12).

We now discuss (with minimum detail) the concept of joint probabilities. Two events are said to be *independent* if the occurrence of one in no way effects the other. On the other hand, two events are said to be *mutually exclusive* if the occurrence of either one precludes the occurrence of the other.

Complementary probability $p(\bar{A})$

If $p(A)$ is the probability that event A occurs, then the probability that event A does not occur is called the complementary probability and is denoted as $p(\bar{A})$. Since it is certain that an event A will either happen or not happen we can write

$$p(A) + p(\bar{A}) = 1.$$

Conditional Probability $p(A/B)$

The probability that event A will occur, given the fact that event B has already occurred, is called conditional probability and is denoted as $p(A/B)$. For independent events the result of event B in no way affects the result of event A and, therefore,

$$p(A/B) = p(A) \quad \text{(independent events)}.$$

However, for mutually exclusive events, we know that if event B has occurred, then the occurrence of event A is impossible. Thus

$$p(A/B) = 0 \quad \text{(mutually exclusive events)}.$$

Probability of both A and B $p(A \cap B)$

If $p(A)$ is the probability that event A will occur and $p(B)$ is the probability that event B will occur, then the probability that *both* A and B will occur is denoted as $p(A \cap B)$ and is described by the formula

$$(11.8) \qquad p(A \cap B) = p(A)p(B/A) = p(B)p(A/B).$$

Again, for mutually exclusive events, $p(A/B) = p(B/A) = 0$ thus

$$p(A \cap B) = 0 \quad \text{(mutually exclusive events)}.$$

For independent events, $p(A/B) = p(A)$ and $p(B/A) = p(B)$. In this case Equation (11.8) becomes

$$(11.9) \qquad p(A \cap B) = p(A)p(B) \quad \text{(independent events)}.$$

Probability of Either A or B $p(A \cup B)$

If $p(A)$ is the probability that event A will occur and $p(B)$ is the probability that event B will occur, then the probability that either A or B will occur is denoted as $p(A \cup B)$ as is described by the formula

$$(11.10) \qquad p(A \cup B) = p(A) + p(B) - p(A \cap B).$$

Once again, for mutually exclusive events $p(A \cap B) = 0$ and, therefore, Equation (11.10) becomes

$$(11.11) \quad p(A \cup B) = p(A) + p(B) \quad \text{(mutually exclusive events)}.$$

For independent events we combine Equations (11.9) and (11.10) to obtain

$$(11.12) \quad p(A \cup B) = p(A) + p(B) - p(A)p(B) \quad \text{(independent events)}.$$

We now consider the idea of a probability distribution, which is simply a frequency distribution whose values x_j are the possible outcomes from an experiment and the f_j terms are the *relative* frequency, or probability, with which the outcomes are expected to occur. For example, in the roll of a single die there are six possible outcomes ranging from one to six and the probability of occurrence of each outcome is easily determined to be $\frac{1}{6}$. Thus our probability distribution is given as

x_j	f_j
1	$\frac{1}{6}$
2	$\frac{1}{6}$
3	$\frac{1}{6}$
4	$\frac{1}{6}$
5	$\frac{1}{6}$
6	$\frac{1}{6}$

Perhaps a more interesting example is obtained by the roll of two dice in which the events are defined to be the sum of the dots showing. From Figure 11.5 we see that the possible outcomes range from $x_1 = 2$ to $x_{11} = 12$. Also using this figure we can construct the probability distribution given in Table 11.3. Shown in Figure 11.6 is a plot of this probability distribution. Note that in this plot we let the values of x_j range from 1 to 13 in which case $p(1) = p(13) = 0$.

Inasmuch as probability distributions are simply special cases of frequency distributions, we are able to characterize them in terms of a mean and standard deviation (or variance). In probability theory the mean, or average, is often called the *expected value*.

Given a probability distribution we are able to make a "best guess" as to the outcome of an experiment. That is, a probability distribution provides us with the likelihood of occurrence of all events in an experiment and, therefore, we are able to use this information to "predict" the outcome. For example, suppose we are asked to guess the total number of spots showing when two dice are rolled simultaneously. From the probability distribution in Table 11.3 or Figure 11.6 we can see that the most likely event to occur is that of showing seven. Thus this should be our guess and, in the long run, we can expect to guess the correct answer $\frac{1}{6}$ of the time.

**TABLE 11.3.   Probability Distribution
for Example of Roll of Two Die**

x_j	f_j
2	$\frac{1}{36}$
3	$\frac{2}{36}$
4	$\frac{3}{36}$
5	$\frac{4}{36}$
6	$\frac{5}{36}$
7	$\frac{6}{36}$
8	$\frac{5}{36}$
9	$\frac{4}{36}$
10	$\frac{3}{36}$
11	$\frac{2}{36}$
12	$\frac{1}{36}$

As another illustration, let us return to our original experiment with the Fourier analysis class summarized as the frequency distribution listed in Table 11.2 and graphed in Figure 11.1. Suppose that we are asked to randomly select one student from this class and guess how much money he or she has. As can be seen from Figure 11.1 the amount that occurs most often is $10 (3 times out of 50) and, therefore, this should be our guess.

In the absence of (or instead of) a probability distribution, what method should we use to determine a best guess as to the outcome of an experiment? Commonly, what is done is to use the computed average, or mean, of the data values. For example, if we compute the average value of our two-dice experiment (Figure 11.5), we obtain $\mu = 7$ which is exactly the answer that we obtained before. However, when this technique is used on the data in Table 11.2 (Fourier Analysis class experiment), we obtain a guess of $24.76 (rounded up to $25). Thus we see that in one case this average value technique agrees with the probability distribution best guess and in the other case it does not. In general, the performance of this technique depends upon many factors such as the symmetry, scatter, and density of the data values. Heuristically, we argue that if the data values are closely grouped around the mean value, then we can expect "reasonably good' results using the average as our best guess. If they are widely dispersed, we can not expect such good results. Inasmuch as the standard deviation, or variance, of a data set is a measure of dispersion, we see that the smaller the standard deviation, the more accurate the average value as a best guess. That is, we can use the computed mean as our best guess and the computed standard deviation as a relative measure of the accuracy of our guess.

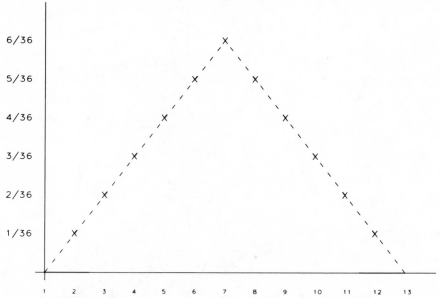

FIGURE 11.6. Graph of a probability distribution of a roll of two dice.

Up to this point, when we constructed our probability distributions, we implicitly assumed that the data values x_j were known exactly. Certainly this is the case when they are the result of a discrete experiment such as the roll of a die. However, more often than not, the values are not known exactly. For example, you may recall that the data presented in Table 11.1 has been "rounded off" to the nearest dollar and thus the data point $22 could actually be any value between $21.50 and $22.49. The more general description of a probability distribution makes use of class intervals. That is, instead of considering the probability $f(\xi_j)$ of an experiment's outcome being exactly equal to ξ_j, we consider $f(\xi_j)$ to be the probability that the outcome is somewhere within an interval (x_j, x_{j+1}). Associated with this interval we define a *probability density function* $p(\xi_j)$ as

(11.13)
$$p(\xi_j) = \frac{f(\xi_j)}{x_{j+1} - x_j} = \frac{f(\xi_j)}{\Delta_j x}.$$

This equation is easily rearranged to read

(11.14)
$$f(\xi_j) = p(\xi_j)\Delta_j x,$$

which expresses the probability $f(\xi_j)$ as the product of the probability density function $p(\xi_j)$ and the class length $\Delta_j x$. Shown in Figure 11.7 is a schematic

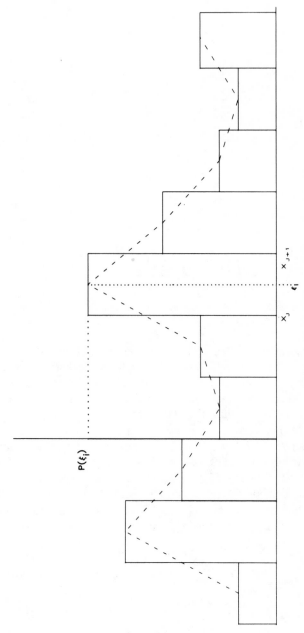

FIGURE 11.7. Schematic diagram of probability density function.

diagram of a probability density function and as can be appreciated from this figure, $p(\xi_j)\Delta_j x$ is the area of the jth rectangle whose height is $p(\xi_j)$ and base is the class interval (x_j, x_{j+1}). In our work we assume that the class intervals all have the same length Δx and also that ξ_j is located at the midpoint of the jth interval (x_j, x_{j+1}). The line segments obtained by connecting the points $[\xi_j, p(\xi_j)]$ provide a graphical representation of the probability density distribution (dashed line in Figure 11.7).

The *cumulative probability density distribution* is defined as

$$(11.15) \qquad P(x_j) = \sum_{i=1}^{j} f(\xi_i) = \sum_{i=1}^{j} p(\xi_i)\,\Delta x.$$

Therefore, based on our previous comments we see that $P(x_j)$ is the probability that the outcome lies between x_1 and x_{j+1}. Similarly,

$$P(x_j) - P(x_k) = \sum_{i=k}^{j} p(\xi_i)\,\Delta x$$

is the probability that the outcome is between x_k and $x_{j+1}(j \geqslant k)$.

It is worthwhile to recall the fact that $p(\xi_i)\,\Delta x$ is a relative frequency and, therefore,

$$(11.16) \qquad P(x_N) = \sum_{i=1}^{N} p(\xi_i)\,\Delta x = 1.$$

The average value and variance of a probability density distribution are given as

$$(11.17) \qquad \mu = \sum_{j=1}^{N} \xi_j p(\xi_j)\,\Delta x,$$

and

$$(11.18) \qquad \sigma^2 = \sum_{j=1}^{N} (\mu - \xi_j)^2 p(\xi_j)\,\Delta x,$$

respectively. These two equations follow directly from Equations (11.2), (11.4), and (11.16).

Up to the point we have been dealing with discrete probability distributions. However, there are many times when the outcome of an event can take on any value over a continuum. For example, we may be observing the output of a physical component which is given as a voltage or displacement. In this case we can still obtain a probability density distribution using class intervals that would be dependent upon both the range and resolution of our observations. For example, let us suppose that we measure the output of a particular component that ranges between -1000 and $+1000$ volts. Also, let us assume that the instrument with which we are making the measurements is graduated in 10-volt increments. Thus we would obtain 200 class intervals each of width 10 volts.

In the theoretical limit of infinitesimal resolution and infinite range we see that the class width Δx approaches zero and the number of intervals become infinite. In this case, the probability density distribution $p(\xi_j)$ becomes a (piecewise) continuous function $p(x)$ over the interval $(-\infty, \infty)$. Also in this theoretical limit Equation (11.15) becomes

$$(11.19) \qquad P(X) = \int_{-\infty}^{X} p(x)\, dx,$$

which is the probability that the outcome lies within the interval $(-\infty, X)$. Furthermore, the probability that the outcome lies within the interval $[A, B]$ is given as

$$(11.20) \quad P(B) - P(A) = \int_{-\infty}^{B} p(x)\, dx - \int_{-\infty}^{A} p(x)\, dx = \int_{A}^{B} p(x)\, dx.$$

The mean and variance of a continuous probability distribution are obtained (as a limit) from Equations (11.17) and (11.18):

$$(11.21) \qquad \mu = \int_{-\infty}^{\infty} x p(x)\, dx,$$

and

$$(11.22) \qquad \sigma^2 = \int_{-\infty}^{\infty} (\mu - x)^2 p(x)\, dx.$$

REVIEW OF CORRELATION THEORY

In this section we give a brief review of correlation analysis. Our attention is limited to those specific areas that are useful to us later in the chapter. The primary purpose of correlation analysis is to measure the strength of a relationship between variables. To illustrate this let us consider the three sets of ordered pairs listed in Table 11.4. We may consider each of the these sets to define a relation. However, a closer examination of the data reveals the fact that the first two relations (S_1 and S_2) are defined using a rigid mathematical

TABLE 11.4. Sets of Ordered Pairs Defining a Relation

S_1			S_2			S_3	
x	y		x	y		x	y
1	2		1	1		1	3
2	4		5	25		6	5
4	8		8	64		7	2
10	20		10	100		9	6
3	6		3	9		8	1
9	18		9	81		2	1
7	14		6	36		4	2
8	16		7	49		10	6
5	10		4	16		3	7
6	12		2	4		5	6

rule, whereas S_3 is not. This becomes somewhat more obvious if we construct scatter diagrams of these data as illustrated in Figure 11.8. From these diagrams it is at least intuitively obvious that the data pairs in sets S_1 and S_2 are strongly related because they seem to lie on a curve. On the other hand, there is no obvious curve that can be constructed through the data in set S_3 and, therefore, we say that these data is weakly related or weakly correlated. This approach to the problem is reasonably satisfactory when the relations are as obvious as those presented in Figure 11.8. However, suppose we have data such as that shown in Figure 11.9. In this case there does seem to be a trend

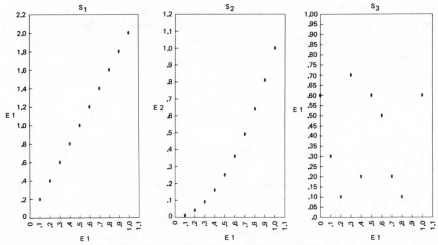

FIGURE 11.8. Scatter diagrams of data in Table 11.4.

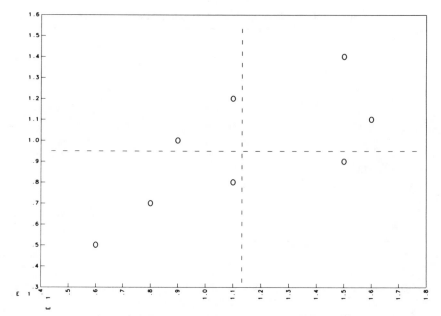

FIGURE 11.9. Scatter diagram and z-scores.

for the y values to increase as the x values increase, but there is certainly no obvious curve upon which all the data values fall. In an effort to be more precise in our discussion of the strength of a relation we offer the following definition. *We say that a set of data pairs (x, y) are perfectly correlated, or related, if there exists a mathematical equation or formula that maps out a curve upon which all these data values lie.* Let us denote this equation as $y = f(x)$. Only under the most ideal circumstances can an equation be found such that every data pair lies on its graph. The problem, therefore, becomes one of finding the "best curve" (or equation) that represents the data. We measure this best curve in terms of its least squares sum:

$$\varepsilon = \sum_{i=1}^{N} \left[f(x_i) - y_i \right]^2.$$

When the data are perfectly correlated, each pair (x_i, y_i) lies on the curve [i.e., $y_i = f(x_i)$] and, consequently, $\varepsilon = 0$. As the data deviate from this curve, the value of ε increases. Thus we now have a quantitative measure of the strength of a relation.

In our presentation so far we have tactfully assumed that we knew the form of the equation $y = f(x)$. That is, we knew if it was a straight line, a parabola, a Gaussian, a Lorentzian, and so on. Actually, the types of curves that we can use are limited only by our imagination. However, the proper choice is usually dictated by the origin of the data and a reasonable amount of good judgment (or luck). We now limit our attention to the case when the curve is a straight line:

$$(11.23) \qquad y = mx + b.$$

This is known as *linear correlation analysis* and it defines the strength of a relation in terms of the data dispersion about a straight line. In this case we optimize the values of m and b to provide the best fit to the data. That is, we find the values of m and b that minimize the error measure given by

$$(11.24) \qquad \varepsilon = \sum_{i=1}^{N} \left[y_i - (mx_i + b) \right]^2.$$

To determine the best fit parameters m and b we must minimize the expression for ε in terms of these parameters. That is, we must calculate $\partial \varepsilon / \partial m$ and $\partial \varepsilon / \partial b$. When we do this and solve the resulting expressions we obtain

$$(11.25) \qquad m = r \frac{\sigma_x}{\sigma_y},$$

$$b = \mu_y - m\mu_x,$$

where μ_x and σ_x are the mean and standard deviation of the distribution of x values and μ_y and σ_y are those for the y values. r is known as *Pearson's r coefficient* and is defined in terms of the z-scores of the two distributions:

$$(11.26) \qquad r = \sum_{i=1}^{N} \frac{z_i(x) z_i(y)}{N},$$

where $z_i(x)$ and $z_i(y)$ are the z-scores of the x and y distributions, respectively. A little algebraic manipulation results in the formula

$$(11.27) \qquad r = \frac{N(\Sigma x_i y_i) - (\Sigma x_i)(\Sigma y_i)}{\sqrt{N(\Sigma x_i^2) - (\Sigma x_i)^2} \sqrt{N(\Sigma y_i^2) - (\Sigma y_i)^2}}.$$

Even though Equation (11.27) appears more complicated than Equation (11.26), in actual practice it turns out to be quite a bit easier to use.

Pearson's r can be used as a measure of the strength of the (linear) relation between a set of data pairs $\langle (x, y) \rangle$. When $|r| = 1$ there is perfect (linear) correlation, whereas $|r| = 0$ implies no correlation at all. Values of $|r|$ in between these extremes indicate varying degrees of correlation ranging from

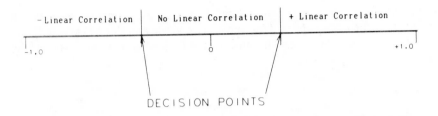

DECISION POINTS

N	Decision point	N	Decision point	N	Decision point	N	Decision point
5	0.878	12	0.576	19	0.456	30	0.361
6	0.811	13	0.553	20	0.444	40	0.312
7	0.754	14	0.532	22	0.423	50	0.279
8	0.707	15	0.514	24	0.404	60	0.254
9	0.666	16	0.497	26	0.388	80	0.220
10	0.632	17	0.482	28	0.374	100	0.196
11	0.602	18	0.468				

FIGURE 11.10. A guide to the use of Pearson's r coefficient.

weakly related to strongly related data. At what point we decide there is sufficient correlation is somewhat nebulous and depends upon the intended use of the data and upon the number of data points. However, the information shown in Figure 11.10 can be used as a guide in making such a decision.

Rather than presenting a rigorous proof of why Pearson's r works as a measure of the strength of a relation, we instead offer the following physical arguments. As already discussed, the z-score of a data value is a measure of its rank or position in a distribution. Therefore, if a set of data pairs $\langle (x, y) \rangle$ is linearly correlated, then it is not unreasonable to expect the data values x and y to have approximately the same z-scores. Let us be more precise and first assume that the data is perfectly correlated:

$$y_i = mx_i + b \qquad (i = 1, \ldots, N).$$

In this case it can easily be shown that $\mu_y = m\mu_x + b$ and $\sigma_y = |m|\sigma_x$. Thus[†]

$$r = \sum_{i=1}^{N} \frac{z_i(x) z_i(y)}{N} = \sum_{i=1}^{N} \frac{(\mu_x - x_i)(\mu_y - y_i)}{\sigma_x \sigma_y N}$$

$$= \frac{\Sigma(\mu_x - x_i)^2 m}{N\sigma_x^2 |m|} = \pm 1.$$

[†] If $m > 0$, then $r = +1$, whereas if $m < 0$, $r = -1$.

Now let us return to the data displayed in Figure 11.9. The dashed lines in this figure mark the locations of the x and y average values. Since z-scores are a measure of the location of a datum value in a distribution with respect to the average value, these dashed lines define the coordinate system for a scatter diagram of the z-scores. For positively correlated data, the $z_i(y)$ values will tend to increase as the $z_i(x)$ values increase. In other words, most of the z-score pairs will lie in the first and third quadrants. Therefore, the resulting products $z_i(x)z_i(y)$ will be (for the most part) positive and their sum, as per Equation (11.26), will produce a positive r value close to 1. For negative correlation, the $z_i(y)$ values will tend to decrease as the $z_i(x)$ values increase. That is, most of the z-score pairs $[z_i(x), z_i(y)]$ will lie in the second and fourth quadrants resulting in an r value close to -1. When there is no correlation at all, the data pairs will be randomly and (on the average) equally distributed throughout all four quadrants resulting in both positive and negative values for the product $z_i(x)z_i(y)$. These products will, therefore, tend to cancel when summed as per Equation (11.26). This results in an r value close to zero.

We end this section with the following example illustrating a simple application of z-scores.

Example

A waterbed store wished to determine if there is a linear correlation between the number of radio commercials broadcast and the number of beds sold. Data were collected over a period of 10 months and summarized as follows.

Month	J	F	M	A	M	J	J	A	S	O
Commercials (x)	27	22	15	35	33	52	35	40	40	40
Beds sold (y)	30	26	25	36	33	36	32	54	50	43

Solution. Our first step in the solution of this problem is to construct the table shown in Figure 11.11. In this form the terms Σx, Σx^2, Σy, Σy^2, and

x	x^2	y	y^2	xy
27	729	30	900	810
22	484	26	676	572
15	225	25	625	375
35	1225	36	1296	1260
33	1089	33	1089	1089
52	2704	36	1296	1872
35	1225	32	1024	1120
40	1600	54	2916	2160
40	1600	50	2500	2000
40	1600	43	1849	1720
339	12481	365	14171	12978

FIGURE 11.11. Linear correlation example problem.

Σxy are readily available. Using Equation (11.27) we determine

$$r = \frac{10(12978) - (339)(365)}{\sqrt{(10)(12481) - (339)^2}\,\sqrt{(10)(14171) - (365)^2}} = 0.66.$$

From Figure 11.10 we see that for 10 data pairs the decision point is 0.632 and, therefore, we can conclude that there is a linear correlation. Also, using the data in Figure 11.11 we find that $\mu_x = 33.9$, $\mu_y = 36.5$, $\sigma_x = 9.94$, and $\sigma_y = 9.21$. Using these values and Equation (11.25) we find that the best least squares straight line has a slope of $m = 0.71$ and y-intercept $b = 12.43$:

$$m = (0.66)\frac{9.94}{9.21} = 0.71,$$

$$b = 36.5 - (0.71)(33.9) = 12.43.$$

This best line is shown in Figure 11.12 along with the original data scatter diagram.

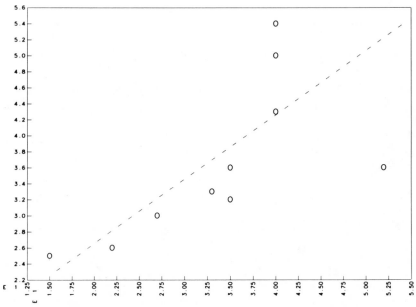

FIGURE 11.12. Linear correlation example problem scatter diagram.

MOMENTS AND THE CENTRAL LIMIT THEOREM

In this section we use the derivative theorems derived in Chapter 3 to obtain both interesting and useful results concerning the mean and variance of a distribution. We also discuss the central limit theorem.

The nth moment of a function (denoted as M_n) is defined as

$$(11.28) \qquad M_n = \int_{-\infty}^{\infty} x^n f(x) \, dx.$$

From Theorem 3.7 we know

$$\frac{1}{(-2\pi i)^n} \frac{d^n F(w)}{dw^n} = \mathscr{F}\left[x^n f(x)\right] = \int_{-\infty}^{\infty} x^n f(x) e^{-2\pi i w x} \, dx$$

and, therefore, if we let $w = 0$, we obtain

$$(11.29) \qquad M_n = \int_{-\infty}^{\infty} x^n f(x) \, dx = \frac{1}{(-2\pi i)^n} \frac{d^n F(0)}{dw^n}.$$

We define the mean μ of a function as

$$(11.30) \qquad \mu = \frac{\displaystyle\int_{-\infty}^{\infty} x f(x) \, dx}{\displaystyle\int_{-\infty}^{\infty} f(x) \, dx} = \frac{M_1}{M_0}.$$

The variance σ^2 of a function is given as

$$(11.31) \qquad \sigma^2 = \frac{\displaystyle\int_{-\infty}^{\infty} (\mu - x)^2 f(x) \, dx}{\displaystyle\int_{-\infty}^{\infty} f(x) \, dx}.$$

Note that if $f(x)$ describes a probability distribution, then

$$\int_{-\infty}^{\infty} f(x) \, dx = 1$$

and the above definitions agree with those of Equations (11.21) and (11.22).

Combining Equations (11.29) and (11.30) we are immediately able to write

$$(11.32) \qquad \mu = \alpha \frac{F'(0)}{F(0)}.$$

where $F'(0) = dF(0)/dw$ and $\alpha = (-1/2\pi i)$. If in a similar way we combine Equations (11.29) and (11.31), we find

$$\sigma^2 = \frac{\alpha^2 F''(0) - 2\mu\alpha F'(0) + \mu^2 F(0)}{F(0)}$$

where $F''(0)$ is the second derivative of $F(w)$ evaluated at $w = 0$. Now making the use of Equation (11.32) in this equation we find

$$\sigma^2 = \alpha^2 \frac{F''(0)}{F(0)} - 2\mu\alpha\frac{\mu}{\alpha} + \mu^2$$

or

(11.33) $$\sigma^2 = \alpha^2 \frac{F''(0)}{F(0)} - \mu^2.$$

We are now in a position to demonstrate the fact that the mean of the convolution of two functions is the sum of the means of the individual functions. We proceed as follows by letting $h(x) = f(x) * g(x)$ and thus by Equation (11.32) we have

$$\mu_h = \alpha\frac{H'(0)}{H(0)}$$

where $H(w)$ is the Fourier transform of $h(x)$. Using the convolution theorem 3.9 we know

$$H(w) = F(w)G(w),$$

from which it follows

$$H'(w) = F'(w)G(w) + F(w)G'(w).$$

Thus

$$\mu_h = \alpha\frac{F'(0)G(0) + F(0)G'(0)}{F(0)G(0)} = \alpha\frac{F'(0)}{F(0)} + \alpha\frac{G'(0)}{G(0)}$$

or

(11.34) $$\mu_h = \mu_f + \mu_g.$$

Similar remarks hold true for the variance of the convolution of two functions.

To demonstrate this we begin with Equation (11.33):

$$\sigma_h^2 = \alpha^2 \frac{H''(0)}{H(0)} - \mu_h^2$$

and use a few basic operations from calculus to show

$$H''(w) = F''(w)G(w) + 2F'(w)G'(w) + F(w)G''(w).$$

When these two equations are combined with Equation (11.34) we obtain

$$\sigma_h^2 = \alpha^2 \frac{F''(0)G(0) + 2F'(0)G'(0) + F(0)G''(0)}{F(0)G(0)} - (\mu_f + \mu_g)^2$$

or

$$\sigma_h^2 = \alpha^2 \frac{F''(0)}{F(0)} + 2\alpha^2 \frac{F'(0)}{F(0)}\frac{G'(0)}{G(0)} + \alpha^2 \frac{G''(0)}{G(0)} - \mu_f^2 - 2\mu_f\mu_g - \mu_g^2.$$

Now making use of Equation (11.32) we find

$$(11.35) \qquad \sigma_h^2 = \alpha^2 \frac{F''(0)}{F(0)} - \mu_f^2 + \alpha^2 \frac{G''(0)}{G(0)} - \mu_g^2 = \sigma_f^2 + \sigma_g^2.$$

We now consider a function $f(x)$ which is defined as the convolution product of n other functions:

$$(11.36) \qquad f(x) = f_1(x) * f_2(x) * f_3(x) * \cdots * f_n(x).$$

Making use of the fact that the convolution product is associative and repeated application of Equations (11.34) and (11.35), we can show that the mean μ and variance σ^2 of $f(x)$ are given as

$$(11.37) \qquad \mu = \mu_1 + \mu_2 + \mu_3 + \cdots + \mu_n.$$

$$(11.38) \qquad \sigma^2 = \sigma_1^2 + \sigma_2^2 + \sigma_3^2 + \cdots + \sigma_n^2,$$

where μ_i and σ_i^2 are the mean and variance of $f_i(x)$ $(i = 1,\ldots, n)$. Using the convolution theorem 3.9 we have

$$F(w) = F_1(w)F_2(w)F_3(w) \cdots F_n(w),$$

and thus

$$(11.39) \qquad \int_{-\infty}^{\infty} f(x)\, dx = F(0) = F_1(0)F_2(0)F_3(0) \cdots F_n(0).$$

Although Equations (11.37) and (11.38) are very revealing and interesting in themselves, there is another property of functions defined as per Equation (11.36) which is quite remarkable. Namely, that after a sufficient number of terms, and under proper conditions, $f(x)$ will take on the shape of a Gaussian distribution. These proper conditions are rigorously spelled out by the central limit theorem (of which there are several versions). Unfortunately, to thoroughly comprehend this theorem we would require a rather sophisticated knowledge of both the mathematical theory of statistics and the theory of higher analysis. For our purposes, it suffices to simply state it:

Central Limit Theorem. If we form the sequence of distributions defined as

$$s_n(x) = f_1(x) * f_2(x) * \cdots * f_n(x),$$

then under *reasonably broad conditions* [placed on $f_i(x)$, $i = 1, \ldots, n$], in the limit as n approaches infinity, the distribution $s_n(x)$ will approach that of a Gaussian whose mean and variance are the sums of the means and variances of the individual distributions $f_i(x)$.

Although the term "reasonably broad" is vague (at best), some appreciation of the conditions can be obtained by considering the preceding equation in the frequency domain:

$$S_n(w) = F_1(w) F_2(w) \cdots F_n(w).$$

Since (in the limit) $s_n(x)$ approaches a Gaussian, then so too must its transform $S_n(w)$ [see Equation (3.16)]. Therefore, some form of the central limit theorem must also apply to a sequence defined (or generated) by the product of the individual distributions. That is, if in the statement of the theorem, we replace the convolution product by the algebraic product, we can obtain the same conclusion (*Note*: the means and variances will not necessarily add in this case). Since multiplication is much easier to "get a handle on" than convolution, we can often gain some insight into these "reasonably broad conditions" by considering the other domain. For example, if the generating distributions for the sequence $S_n(w)$ are all equal and given by

$$F_i(w) = \frac{\delta(w + a) + \delta(w - a)}{2} \qquad i = 1, \ldots, n,$$

then clearly $S_n(w)$ can never evolve into a Gaussian. Inverse transforming these generating sequences, we can conclude that if the generating distributions are given as

$$f_i(x) = \cos 2\pi a x$$

then the sequence $s_n(x)$ formed as the nth convolution product of $f_i(x)$ will never evolve into the Gaussian profile.

Next we consider examples of distributions that do satisfy the conditions of the central limit theorem. Let us consider the sequence $s_n(x)$ defined as

$$s_n(x) = f_1(x) * f_2(x) * \cdots * f_n(x),$$

where $f_i(x)$ are all equal pulse functions of half-width 1. In Figure 11.13 we show the first two (digitally obtained) convolution products of this sequence and we find (*Note*: $\mu_i = 0$ and $\sigma_i^2 = \frac{2}{3}$)

$$\mu \approx 0 \quad \text{and} \quad \sigma^2 \approx \sigma_1^2 + \sigma_1^2 + \sigma_1^2 = \frac{6}{3} = 2.$$

Next we consider an example in which $s_n(x)$ is defined as the convolution of equal one-sided exponentials. That is,

$$s_n(x) = g_1(x) * g_2(x) * \cdots * g_n(x),$$

where

$$g_i(x) = \begin{cases} 0, & x < 0 \\ e^{-x}, & x \geqslant 0. \end{cases} \quad i = 1, \dots, n$$

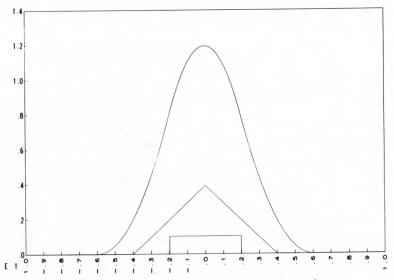

FIGURE 11.13. Evolution of the pulse convolution product.

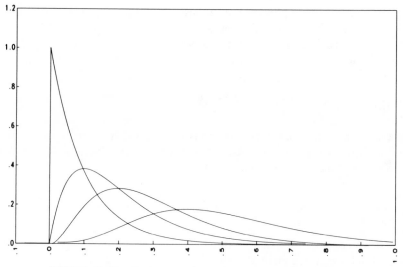

FIGURE 11.14.   Evolution of the exponential convolution product.

It can easily be shown that

$$\mu = \int_0^\infty e^{-x}\,dx = 1 \quad \text{and} \quad \sigma^2 = \int_0^\infty (1-x)^2 e^{-x}\,dx = 1.$$

In Figure 11.14 we show the first three convolution products from which it can clearly be seen that $s_n(x)$ is indeed evolving into a Gaussian profile. Note in this figure that $\mu \approx 4$ and $\sigma^2 \approx 4$.

CONVOLUTION AND PROBABILITY DISTRIBUTIONS

In a previous section we discussed the concept of joint probabilities. That is, given the probabilities of two single events A and B, we were able to determine the probability of the occurrence of both A and B or the occurrence of either A or B. In this section we show how (under proper conditions) the concept of convolution can be used to combine two or more probability distributions.

We begin our discussion by returning to one of our previous examples in which two dice were rolled simultaneously and we were interested in the total number of spots showing. This is shown schematically in Figure 11.5. Using this chart and brute force methods, we were able to construct the probability distribution shown in Figure 11.6. Let us now generate this distribution in a more general way. Let $p_1(i)$ and $p_2(i)$ be the individual probability distributions for the roll of each die:

$$p_1(i) = p_2(i) = \begin{cases} \frac{1}{6}, & i \in [1,6] \\ 0, & \text{otherwise} \end{cases}$$

These are shown in Figure 11.15. We now wish to determine the probability that the sum of the two dice is equal to the number k:

$$k = i_1 + i_2 \quad \text{or} \quad i_2 = k - i_1.$$

That is, if the roll of the first die yields i_1, then the roll of the second must be equal to $k - i_1$. The probability of occurrence of this particular combination (of independent events) is given as the product of the individual probabilities (Equation 11.9); that is, $p_1(i)p_2(k - i)$. (*Note*: We use i for i_1 from here on.) Since there are six possible values for i we can write

$$p(k) = \sum_{i=1}^{6} p_1(i)p_2(k - i), \qquad k = 2,\ldots,12.$$

Inasmuch as both $p_1(i)$ and $p_2(i)$ are equal to zero if i is outside the interval $[1, 6]$, we write this equation with new limits (*Note*: $N > 12$):

$$(11.40) \qquad p(k) = \sum_{i=0}^{N-1} p_1(i)p_2(k - i) \qquad k \in [0, N - 1].$$

This equation may be considered as the discrete convolution of $\{p_1(k)\}$ and $\{p_2(k)\}$. Performing the summation we obtain the same probability distribution as before, shown in Figure 11.6.

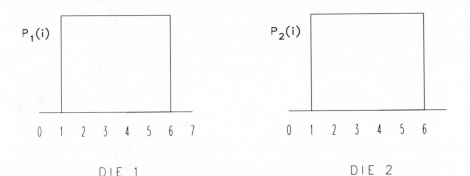

DIE 1   DIE 2

FIGURE 11.15. Probability distribution for two single die.

The extension of this concept to continuous distributions is straightforward. Shown in Figure 11.16 are two probability distributions $p_1(\xi_1)$ and $p_2(\xi_2)$ which give the relative frequency of occurrence of the values ξ_1 and ξ_2, respectively. Now let us suppose that we are going to combine these two distributions by choosing values from each and adding them together. The problem at hand is to determine the resulting probability distribution of these sums $x = \xi_1 + \xi_2$. Just as we did in the discrete case, we begin by rewriting this equation as $\xi_2 = x - \xi_1$. Next we assume that we choose the value ξ_1 from the first distribution. Then if the total value is to be equal to x, we must choose the second value to be $\xi_2 = x - \xi_1$. The probability of this happening is

$$p_1(\xi_1)p_2(x - \xi_1).$$

Now "summing" over all possible values of ξ_1 we find

$$p(x) = \int_{-\infty}^{\infty} p_1(\xi_1)p_2(x - \xi_1)\,d\xi_1$$

or

(11.41) $$p(x) = p_1(x) * p_2(x).$$

From the previous section we know that the mean and variance of this new distribution are simply the sums of the individual means and variances of the function we convolved.

Using the central limit theorem we know that if we repeat this type of procedure several times (i.e., choose values from several distributions and add them together), then we will arrive at a distribution $p(x)$ which "strongly resembles" a Gaussian, or normal, profile whose mean and standard deviation are the sums of the individual means and standard deviations.

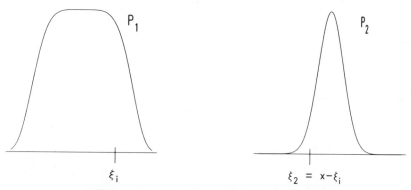

FIGURE 11.16. Continuous probability distributions.

AUTOCORRELATION AND RANDOM SIGNALS

In this section we give a heuristic discussion of how correlation analysis can be used as a test for the randomness of a function. The purpose of this section is to present a physical interpretation of autocorrelation when applied to random or stochastic signals. A more rigorous description is given in the following sections.

Let $r(t)$ be a random function or signal such as that shown in Figure 11.17. For any given t, the value $r(t)$ cannot be described by a mathematical formula. In other words, it must be expressed in terms of an expected value and standard deviation (we have more to say about this later). Thus for a random function we do not expect an analytical relationship between any two values $r(t_1)$ and $r(t_2)$. To obtain a better appreciation for this, let us briefly digress and discuss deterministic, or analytical, functions. The problem at hand is to determine the relationship between any two values $f(t_1)$ and $f(t_2)$ or, if we let $t_1 = t$ and $t_2 = t + \tau$, the relationship between any two values $f(t)$ and $f(t + \tau)$. We learned in a previous section that we could examine the relationship or correlation between two functions (or sets) by plotting them as a scatter diagram. Thus if we let $x(t) = f(t)$ and $y(t) = f(t + \tau)$ and plot them against each other, then we can obtain a visual representation of the relationship between them. We call these plots *autocorrelation graphs*. For example, if

$$f(t) = e^{-at},$$

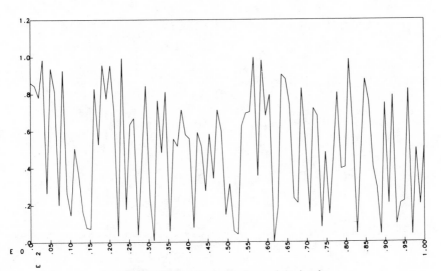

FIGURE 11.17. Example of a random signal.

then

$$x(t) = f(t) = e^{-at},$$

and

$$y(t) = f(t + \tau) = e^{-a(t+\tau)} = e^{-at}e^{-a\tau} = e^{-a\tau}x(t).$$

Certainly, for $\tau = 0$ we have perfect positive correlation; that is, $x(t) = y(t)$. In Figure 11.18 we show the autocorrelation graphs of $f(t) = e^{-at}$ for several (increasing) values of τ.

As a second, and perhaps more interesting, example let us consider the function

$$f(t) = \sin t.$$

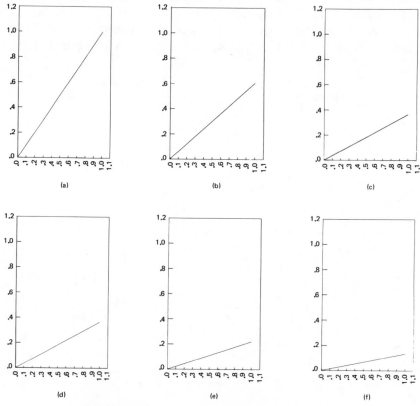

FIGURE 11.18. Autocorrelation graphs of $f(t) = e^{-at}$.

In this case we let $x(t) = \sin t$ and $y(t) = \sin(t + \tau)$, or

$$y(t) = \sin t \cos \tau + \cos t \sin \tau,$$

$$= x(t)\cos \tau + \frac{dx(t)}{dt}\sin \tau.$$

Clearly, $y(t)$ and $x(t)$ are related, but not in a simple fashion as in our previous example. Again for $\tau = 0$ we have perfect correlation; that is, $y(t) = x(t)$. In Figure 11.19 we show the autocorrelation graphs for several different (increasing) values of τ.

Now, let us return to our original task of examining the relationship between any two values of a random function $r(t)$ and $r(t + \tau)$. We do this by considering the function at equal increments in time and constructing scatter diagrams. That is, we let

$$x_k = r(k\Delta t) \quad \text{and} \quad y_k = r(k\Delta t + \tau)$$

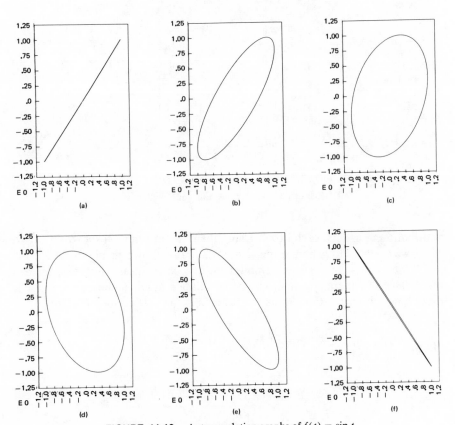

FIGURE 11.19. Autocorrelation graphs of $f(t) = \sin t$.

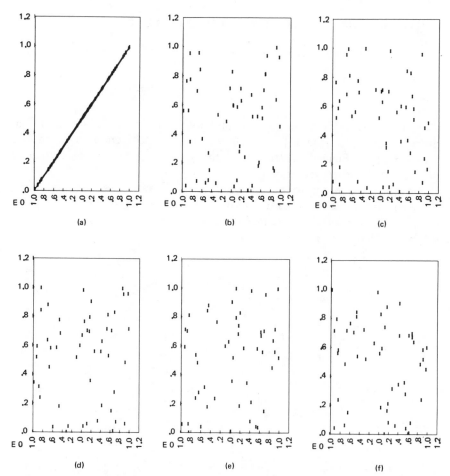

FIGURE 11.20.   Scatter diagrams for a random signal.

and plot them against each other. Using the random function shown in Figure 11.17 we are able to generate the scatter diagrams shown in Figure 11.20. For $\tau = 0$ we have $x_k = y_k$ which implies perfect correlation. However, for any other value of τ there appears to be no correlation at all. To be more precise in describing this correlation, let us use Pearson's r score [Equation (11.26)]:

$$(11.42) \qquad P_r(\tau) = \sum_{k=1}^{N} \frac{1}{N} \left(\frac{x_k - \mu_x}{\sigma_x} \right) \left(\frac{y_k - \mu_y}{\sigma_y} \right)$$

$$= \sum_{k=1}^{N} \frac{1}{N} \frac{r(k\Delta t) - \mu_x}{\sigma_x} \frac{r(k\Delta t + \tau) - \mu_y}{\sigma_y}.$$

The r-scores for the scatter diagrams of Figure 11.20 are determined to be: (a) 1.000, (b) 0.077, (c) -0.146, (d) 0.0443, (e) 0.209, and (f) -0.179. As can be appreciated by these calculations, there is (essentially) no correlation unless $\tau = 0$.

To simplify Equation (11.42), let us make the assumption that $\mu_x = \mu_y = \mu$ and $\sigma_x = \sigma_y = \sigma$.[†] When this is accomplished, we can rewrite Equation (11.42) as

$$(11.43) \qquad P_r(\tau) = \sum_{k=1}^{N} \frac{r(k\Delta t)r(k\Delta t + \tau)}{N\sigma^2} - \frac{\mu^2}{\sigma^2}.$$

For the ideal case of a completely random function we have already indicated that

$$P_r(\tau) = \begin{cases} 1, & \tau = 0 \\ 0, & \text{otherwise,} \end{cases}$$

or

$$P_r(\tau) = \delta(\tau) \quad \text{(random function)}.$$

In actual practice it is usually more convenient to calculate the summation term in Equation (11.43):

$$(11.44) \qquad R(\tau) = \sum_{k=1}^{N} r(k\Delta t)r(k\Delta t + \tau).$$

Thus let us rewrite Equation (11.43) as

$$(11.45) \qquad R(\tau) = N\sigma^2 P_r(\tau) + N\mu^2.$$

Again, for a completely random function, Equation (11.45) tells us that $R(\tau)$ is a delta function of strength $N\sigma^2$ added to a constant value $N\mu^2$. In the limit as $\Delta t \to 0$ and $N \to 0$, $r(k\Delta t)$ becomes a continuous random signal and Equation (11.44) becomes

$$(11.46) \qquad R(\tau) = \int_{-\infty}^{\infty} r(t)r(t + \tau)\, dt,$$

which is simply the autocorrelation function of $r(t)$ with itself.

[†] This assumption turns out to be quite reasonable if the signals are long in duration compared to the value of the offset τ. Also, we consider the limit as these sampled signals become continuous and infinite in duration.

We can again argue that for a completely random continuous signal, its autocorrelation $R(\tau)$ will be an impulse or delta function (added to a constant). Thus we see that the autocorrelation integral can be used as a test to determine if a signal is completely random. On the other end of the spectrum, we know that if $r(t)$ is a completely deterministic signal, then its autocorrelation will also be deterministic. For example, if $r(t)$ is a pulse function, then $R(\tau)$ is a triangle function. Real world signals are usually somewhere in between these extreme cases and, consequently, their autocorrelation is somewhere between an impulse and a deterministic function. Admittedly, this is a rather nebulous statement and the ability to judge the randomness of a signal based on its autocorrelation is a talent that must be acquired through practice.

In this section we have illustrated how autocorrelation can be used to characterize a random signal. The purpose of this section was to give a physical or intuitive description of the subject and, therefore, it was presented with a minimum (or perhaps lack) of rigor. In the next section we are somewhat more formal in our discussions.

ENSEMBLES AND EXPECTED VALUES OF A RANDOM SIGNAL

We have already noted in the previous sections that random signals cannot be described by a mathematical formula. That is, they cannot be described in terms of their pointwise properties and thus when we discuss them we must consider their overall or statistical properties. This statistical description of a random signal can be accomplished in either of two ways. One approach chooses to define the average value of a random signal in terms of the equation

$$(11.47) \qquad \langle r(t) \rangle = \lim_{T \to \infty} \frac{1}{2T} \int_{-T}^{T} r(t)\, dt.$$

The other approach, and the one that we take in this section, is to consider a random signal to be a member of an ensemble of such signals. For example, if each day for a year we were to record the wind velocity at a particular location, then we would obtain an ensemble of 365 recordings of a random signal. The individual members of an ensemble are called sample signals or sample functions. In Figure 11.21 we show three sample signals from an ensemble. In this figure we indicate the location of a particular value of the independent variable t. For this value t we can calculate the average value of the random signal over the ensemble:

$$(11.48) \qquad \mu(t) = \sum_{i=1}^{N} \frac{r_i(t)}{N}.$$

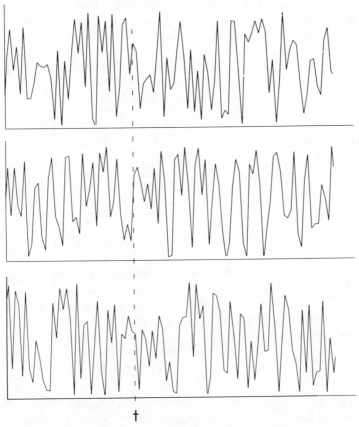

FIGURE 11.21. Three recordings of a random signal.

In our work we are interested in the situation when the ensemble is very large or (for all practical purposes) infinite. In this case we are motivated to define the expected value of the signal $r(t)$ [denoted as $\eta_r(t)$ or $E\{r(t)\}$] as

$$(11.49) \qquad \eta_r(t) = E\{r(t)\} = \lim_{N \to \infty} \frac{1}{N} \sum_{i=1}^{N} r_i(t).$$

Let us now consider a few interesting properties of this expected value function. First of all, it maps any value t to $\eta_r(t)$ as per Equation (11.49) and, therefore, it is a deterministic function. Also, when we set up the definition of $\eta_r(t)$ we implied that we were only dealing with an ensemble generated by a random signal. However, if the ensemble is, in fact, generated by a deterministic signal, we have $r_i(t) = r(t)$ (for all i). In this case we obtain $\eta_r(t) = r(t)$.

We now show that the expected value function is linear. To do so let $x(t)$, $y(t)$, and $z(t)$ be three different signals that generate three separate ensembles. Also let a, b, and c be any constants. Then we must demonstrate that

$$E\{ax(t) + by(t) + cz(t)\} = aE\{x(t)\} + bE\{y(t)\} + cE\{z(t)\}.$$

This turns out to be a rather straightforward task:

$$E\{ax(t) + by(t) + cz(t)\} = \lim_{N \to \infty} \frac{1}{N} \sum_{i=1}^{N} \left[ax_i(t) + by_i(t) + cz_i(t)\right]$$

$$= \lim_{N \to \infty} \left[a \sum_{i=1}^{N} \frac{x_i(t)}{N} + b \sum_{i=1}^{N} \frac{y_i(t)}{N} + c \sum_{i=1}^{N} \frac{z_i(t)}{N}\right]$$

$$= aE\{x(t)\} + bE\{y(t)\} + cE\{z(t)\}.$$

This can be generalized to include an infinite number of random signals.

We now define the crosscorrelation function of two signals based on the concept of ensemble averages. Later in the chapter we compare this to our previous definition (Chapter 3). Given two signals $x(t)$ and $y(t)$, we define their crosscorrelation product [denoted as $R_{xy}(t_1, t_2)$] as

(11.50) $$R_{xy}(t_1, t_2) = E\{x(t_1)y^*(t_2)\}.$$

To evaluate this expected value we first form the product of the signal $x(t)$ evaluated at t_1 and the complex conjugate of the signal $y(t)$ evaluated at t_2 for each member in the ensemble. We then compute the average or expected value of this product:

$$E\{x(t)y^*(t)\} = \lim_{N \to \infty} \sum_{i=1}^{N} \frac{x_i(t_1)y_i^*(t_2)}{N}$$

Similarly, the autocorrelation of a signal $x(t)$ is defined as

(11.51) $$R_{xx}(t_1, t_2) = E\{x(t_1)x^*(t_2)\}.$$

Note that when $t_1 = t_2 = t$ we have

$$R_{xx}(t, t) = E\{x(t)x^*(t)\} = E\{|x(t)|^2\}.$$

$|x(t)|^2$ is an important quantity for physical systems because it is usually measurable. For example, in Chapter 8 we noted that $|\psi(x)|^2$ is the observable quantity and not the optical disturbance function $\psi(x)$ itself.

SYSTEMS AND RANDOM SIGNALS

In this section we consider how a deterministic system responds to an input signal that is stochastic or random. When we say a system is deterministic we mean that its impulse response function (see Chapter 6) is deterministic. Let us denote this impulse response as $h(t)$. As input to this system we use the signal $x(t)$ with expected value $\eta_x(t)$. As we learned in Chapter 6, the resulting output to this system $y(t)$ is given as

$$y(t) = \int_{-\infty}^{\infty} x(\alpha) h(t - \alpha) \, d\alpha.$$

Thus the expected value of this output signal is

$$\eta_y(t) = E\{y(t)\} = E\left\{ \int_{-\infty}^{\infty} x(\alpha) h(t - \alpha) \, d\alpha \right\}.$$

However, because of the previously mentioned linearity property of expected values, we can bring this operation inside the integral:

$$\eta_y(t) = \int_{-\infty}^{\infty} E\{x(\alpha) h(t - \alpha) \, d\alpha\}.$$

Furthermore, inasmuch as the impulse response function $h(t)$ is deterministic, we can remove it from the expected value brackets to obtain

$$\eta_y(t) = \int_{-\infty}^{\infty} E\{x(\alpha)\} h(t - \alpha) \, d\alpha = \int_{-\infty}^{\infty} \eta_x(\alpha) h(t - \alpha) \, d\alpha$$

$$(11.52) \quad \eta_y(t) = \eta_x(t) * h(t).$$

This equation tells us that the expected value of the output signal is simply the convolution of the expected value function of the input signal with the impulse response function of the system. That is, the system operates on the expected value of the input signal to yield the expected value of the output signal.

Suppose that for this system we take the output $y(t)$, crosscorrelate it with the input $x(t)$ to obtain $R_{xy}(t_1, t_2)$, and then use this as an input signal to the system. We now show that when this is done, the resulting output is the autocorrelation of $y(t)$ with itself, that is, $R_{yy}(t_1, t_2)$. To accomplish this we begin with

$$y(t) = \int_{-\infty}^{\infty} x(t - \alpha) h(\alpha) \, d\alpha$$

and multiply both sides by $y^*(t_2)$ to obtain

$$y(t)y^*(t_2) = \int_{-\infty}^{\infty} x(t - \alpha)y^*(t_2)h(\alpha)\, d\alpha.$$

Next we take the expected value of both sides of the preceding equation. This move results in

$$E\{y(t)y^*(t_2)\} = \int_{-\infty}^{\infty} E\{x(t - \alpha)y^*(t_2)\}h(\alpha)\, d\alpha.$$

Now if we treat t_2 as a parameter in this equation we find

$$(11.53) \quad R_{yy}(t, t_2) = \int_{-\infty}^{\infty} R_{xy}(t - \alpha, t_2)h(\alpha)\, d\alpha = R_{xy}(t, t_2) * h(t).$$

Another interesting result can be obtained by considering the conjugate system. Given a system with impulse response $h(t)$, we define its *conjugate system* as that which has an impulse response given as $h^*(t)$. Obviously, a real system is its own conjugate. Using logic similar to that which led us to Equation (11.53), we can also show that if the input to this conjugate system is $R_{xx}(t_1, t)$, then the resulting output is given as $R_{xy}(t_1, t)$:

$$(11.54) \quad R_{xy}(t_1, t) = \int_{-\infty}^{\infty} R_{xx}(t_1, t - \alpha)h^*(\alpha)\, d\alpha = R_{xx}(t_1, t) * h^*(t).$$

In this case the convolution is carried out with respect to t, and t_1 is considered the parameter. The results obtained so far in this section are schematically summarized in Figure 11.22.

It is important to note that the results presented so far in this section apply to deterministic as well as random signals. We now limit our attention to signals that are completely random or incoherent.[†] We say a signal $x(t)$ is completely random if and only if

$$R_{xx}(t_1, t_2) = 0, \qquad t_1 \neq t_2.$$

When $t_1 = t_2$ we have

$$(11.55) \quad R_{xx}(t_1, t_1) = E\{x(t_1), x^*(t_1)\} = E\{|x(t_1)|^2\} = f(t_1).$$

Thus for an incoherent signal we can write

$$(11.56) \qquad R_{xx}(t_1, t_2) = f(t_1)\delta(t_2 - t_1).$$

[†]Completely random signals are also known as white noise.

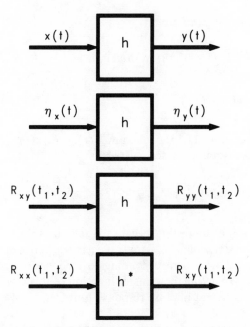

FIGURE 11.22. Schematic summary of some stochastic system results.

We now show that if $h(t)$ defines a system that maps $x(t)$ to $y(t)$, then the system defined by $|h(t)|^2$ will map $E\{|x(t)|^2\}$ to $E\{|y(t)|^2\}$. To demonstrate this fact we begin with the conjugate system Equation (11.54) and use as input the signal given by Equation (11.56):

$$R_{xy}(t_1, t_2) = R_{xx}(t_1, t_2) * h^*(t_2) = f(t_1)\delta(t_2 - t_1) * h^*(t_2).$$

The above convolution is performed over the variable t_2 and results in

$$(11.57) \qquad R_{xy}(t_1, t_2) = f(t_1)h^*(t_2 - t_1).$$

We now use this signal as input to the system described by Equation (11.53) to obtain

$$R_{yy}(t_1, t_2) = R_{xy}(t_1, t_2) * h(t_1) = f(t_1)h^*(t_2 - t_1) * h(t_1).$$

This convolution is performed over the variable t_1. Our results can be more easily appreciated if we write out the equation in its integral form:

$$R_{yy}(t_1, t_2) = \int_{-\infty}^{\infty} f(t_1 - \alpha)h^*(t_2 - t_1 + \alpha)h(\alpha)\,d\alpha.$$

Now let us set $t = t_1 = t_2$ to obtain

$$R_{yy}(t, t) = \int_{-\infty}^{\infty} f(t - \alpha) h^*(\alpha) h(\alpha) \, d\alpha = \int_{-\infty}^{\infty} f(t - \alpha) |h(\alpha)|^2 \, d\alpha$$

$$R_{yy}(t, t) = f(t) * |h(t)|^2.$$

However, by Equation (11.55) we know $f(t) = E\{|x(t)|^2\}$ and $R_{yy}(t, t) = E\{|y(t)|^2\}$. Thus we obtain our desired results:

(11.58) $$E\{|y(t)|^2\} = E\{|x(t)|^2\} * |h(t)^2|.$$

This equation is quite interesting because it tells us that if we know how a system behaves for deterministic signals [i.e., if we know $h(t)$], then we also know how it will behave when responding to incoherent signals [i.e., we can determine $|h(t)^2|$]. We define the transfer function of this system as the Fourier transform of its incoherent impulse response function:

$$\mathscr{F}\left[|h(t)|^2\right] = \mathscr{F}\left[h(t)h^*(t)\right] = \mathscr{F}\left[h(t)\right] * \mathscr{F}\left[h^*(t)\right].$$

However, if we denote $\mathscr{F}[h(t)]$ as $H(w)$, then we obtain

(11.59) $$\mathscr{F}\left[|h(t)|^2\right] = H(w) * H^*(-w).$$

STATIONARY SIGNALS AND THE ERGODICITY PRINCIPLE

In this section we limit our attention to stationary signals that are a special case of random signals. For a signal to be *stationary* it must satisfy two conditions, namely;

1. Its expected value, or mean, must be a constant; that is, $E\{x(t)\} = \eta_x$.
2. Its autocorrelation[†] must only be a function of $t_2 - t_1$, that is, $R_{xx}(t_2 - t_1)$.

For convenience, we let $\tau = t_2 - t_1$. Using Equation (11.51) we have

$$R_{xx}(\tau) = E\{x(t_1)x^*(t_2)\} = E\{x(t)x^*(t + \tau)\}.$$

[†]Whenever the autocorrelation $R_{xx}(\tau)$ is written as a function of only one variable, we implicitly assume that the signal is stationary.

We now demonstrate that if the input to a linear deterministic system is stationary, then so too will be the output. To accomplish this demonstration we must show that the output signal satisfies conditions 1 and 2. In the previous section [Equation (11.52)] we showed

$$\eta_y(t) = \int_{-\infty}^{\infty} \eta_x(\tau)h(t - \tau)\, d\tau.$$

However, inasmuch as $\eta_x(t) = \eta_x$ is a constant, we find

$$\eta_y(t) = \eta_x \int_{-\infty}^{\infty} h(t - \tau)\, d\tau = \text{constant}$$

and thus condition 1 is satisfied. Next we must show that the autocorrelation of the output signal is only a function of $\tau = t_2 - t_1$. For our starting point we refer to Equation (11.54) of the previous section:

$$R_{xy}(?) = R_{xx}(\tau) * h^*(\tau).$$

This equation clearly implies that $R_{xy}(?)$ is only a function of $\tau = t_2 - t_1$. Next we use this result in Equation (11.53) to obtain

$$R_{yy}(?) = R_{xy}(\tau) * h(\tau)$$

which again implies $R_{yy}(?)$ is only a function of $\tau = t_2 - t_1$.

Given a stationary signal $y(t)$, we define its *power spectrum* as the Fourier transform of its autocorrelation, that is, $\mathcal{F}[R_{xx}(\tau)]$. For an incoherent signal we have $R_{xx}(\tau) = K\delta(\tau)$ and, therefore, its power spectrum is a constant.

The ergodicity theorem provides us with the conditions under which the time average of a random signal (as per Equation 11.47) is equal to its expected value of ensemble average (as per Equation 11.49). Although we do not become embroiled in the mathematical details here, it turns out that for stationary random signals it is indeed true that the time averages are equal to the ensemble averages. That is,

$$E\{r(t)\} = \lim_{T \to \infty} \frac{1}{2T} \int_{-\infty}^{\infty} r(t)\, dt = A \int_{-\infty}^{\infty} r(t)\, dt.$$

This fact provides us with the connecting link between this chapter's definition of autocorrelation and that presented in Chapter 3:

$$(11.60) \quad R_{xx}(\tau) = E\{x(t)x^*(t + \tau)\} = A \int_{-\infty}^{\infty} x(\alpha)x^*(\tau + \alpha)\, d\alpha.$$

SUMMARY

In this chapter we discussed how Fourier analysis can be used to help solve problems encountered in the areas of probability and statistics. We also briefly studied how linear deterministic systems behave when responding to random or stochastic signals.

BIBLIOGRAPHY

Bracewell, R., *The Fourier Transform and it Applications*, 2nd ed. McGraw-Hill, New York, 1978.

Papoulis, A., *Systems and Transforms with Applications in Optics*. McGraw-Hill, New York, 1968.

Raven, F. H., *Mathematics of Engineering Systems*. McGraw-Hill, New York, 1966.

Cramer, H., *Mathematical Methods of Statistics*. Princeton University Press, Princeton, N.J. 1963.

Ingram, I. A., *Introductory Statistics*. Cummings Publishing, Menlo Park, Calif., 1974.

INDEX